ENGINEERS' ILLUSTRATED THESAURUS

by
Herbert Herkimer

CHEMICAL PUBLISHING CO., INC.
NEW YORK, N. Y.

Printed in the United States of America

FOREWORD

More than thirty years ago, the author began to follow the advice given by William Kent to his students:

"Every young engineer should compile his own pocket-data book, as he proceeds in study and practice, to suit his particular business."

The author went a step farther and began his scrap-book collection of engineering drawings, sketches, diagrams and abstracts from various domestic and foreign patent-office gazettes and trade catalogs.

At the request of the publishers, he started the work of selecting from this accumulated material data pertaining to mechanical engineering and other engineering fields, and arranging them in a form suitable for publication.

The main object has been to present the maximum number of illustrations; this naturally limited descriptive text to a minimum, to keep the book from growing beyond all practical bounds. Engineering science and practice have developed to such an extent that a detailed analysis of the more than 8000 illustrations in this book would fill many volumes.

The graphic method of describing machine parts and their movement by means of diagrams, line drawings and photographs has been generally accepted as the quickest and most satisfactory. Since the purpose of the book is to emphasize underlying principles and not structural details, such as would appear in a textbook of machine design, the sketches and drawings will present a maximum of useful basic information without confusing irrelevant detail, since the practical engineer and inventor need only an outline of an idea for his inspiration and would probably resent elaborate explanations.

While it may be repeated that this work is not a textbook of machine design, nevertheless typical assembly drawings, examples of American and foreign designs are given. It is a well-known fact that most complicated mechanisms consist merely of combinations of the six fundamental machines:

1. Pulley
2. Wheel and axle
3. Inclined plane
4. Wedge
5. Screw
6. Lever

Yet the field of mechanisms and structures seems almost unlimited, according to the patent-office records and the volume of trade catalogs published monthly.

One claim to originality by an author of a book of this type lies in the novel methods of arrangement and indexing for ready, rapid reference.

The author has adopted a classification similar to that of *Roget's Thesaurus of English Words and Phrases* in which a word is classed according to the idea it intends to convey. In addition, there is a detailed alphabetical index in the back of the book. In using the book, the classification should be consulted first and the index afterward.

As all bodies are either in a state of rest or in motion, accordingly two main divisions are used as follows:

PART I — STATICS

Class I   — Fasteners
Class II   — Adjusting Devices
Class III   — Supports and Structures

PART II — DYNAMICS

Class IV   — Basic Mechanical Movements
Class V   — Elevators, Derricks, Cranes, Conveyors
Class VI   — Transmission of Liquids and Gases
Class VII   — Combustion
Class VIII   — Prime Movers
Class IX   — Transportation
Class X   — Industrial Processes
Class XI   — Electrical Appliances
Class XII   — Comfort Heating Cooling and Air Conditioning

Each class is divided into sections and each section is further subdivided into topics. Each topic is illustrated by drawings or photographs.

Invention and engineering design constitute a peculiar art which cannot be acquired but by long and continued practice. There are some engineers more highly gifted than others, but to all there comes a time when ideas stagnate and the solution is far away. Like 'spirits from the vasty deep they come not when we call.'

To engineers struggling with difficulties, this book should prove of great help in solving their problems.

Although with some machine parts, the name of the manufacturer is given, this does not mean that the author recommends the said manufacturer's products. All illustrations are given as examples only and it is left to the reader to select equivalent products of other manufacturers, if he prefers.

The author extends his thanks to many manufacturers who have so courteously supplied drawings and catalogs. If there is any borrowed matter of importance, the source of which is unknown to the author, he will be grateful for pointing it out to him and he will acknowledge it in a later edition.

Grateful acknowledgment is also due to the author's son, Harold Herkimer, for his assistance in the preparation of the illustrations.

TABLE OF CONTENTS

Contents

Contents

Contents

Contents

Contents

CLASS XI: ELECTRICAL APPLIANCES (Cont'd)
Sections

CLASS XII: COMFORT HEATING, COOLING AND AIR CONDITIONING   520
Sections

INTRODUCTION

Engineers, designers and draftsmen deal with machines. A machine is defined as a combination of parts which is suitable to transmit and modify energy and motion to do the desired work. Another definition of a machine describes it as a device that overcomes resistance at one point by the application of force at some other point.

Energy may be defined as the ability to do work. There are two main types of energy: Potential energy is latent until a change releases it. The energy stored in coal (chemical energy) and changed into heat by burning and the energy of water in a high tank (energy of position) which is released by opening a valve are forms of potential energy. Other examples of potential energy are that of a raised weight, wound spring, and compressed gas. Kinetic energy is the energy of motion. When potential energy is released, it is transformed into kinetic energy. Kinetic energy is sometimes called mechanical energy. Examples of kinetic energy are the electric current, heat, light, energy of expanding gas, working muscles, combining elements, the energy released by atomic fissure, etc.

The law of conservation of energy states that in any isolated system, the total amount of energy is constant. This means that energy can change from one form to another but the total amount of energy will remain the same. This may be expressed mathematically:

Kinetic Energy + Potential Energy = Constant

or

Total Energy Deposited = Work Accomplished +
Energy Lost by Resistance

As the complete energy supplied cannot be converted into useful work, but a certain portion of it is always used to overcome resistance, the idea of a perpetual-motion machine is absurd.

Prime movers are machines which convert energy from a

natural source into mechanical power. Oil engines, gas engines, wind mills, water wheels and turbines, steam engines and internal-combustion engines are examples of prime movers.

Power is the rate of doing work. The standard unit of power is one horse power which is equal to 33,000 pounds lifted 1 foot high in 1 minute.

If the number of foot-pounds done per minute is known, we can express the work in horse-power units by dividing by 33,000. For example:

	Foot-pounds	Horse power
A man raising his own weight vertically during a day of eight hours	4,350	0.1318
A man pushing and pulling at capstan	3,180	0.0963
A man turning a winch	2,700	0.0818
A horse pulling a cart	26,150	0.7924

This shows that a man performs 1/10 to 1/8 of a horse power and a real horse 8/10 of a horse power.

Force is the cause of the acceleration of a body free to move. Its unit is the poundal or the dyne.

Velocity is the rate of movement and is measured in feet or centimeters per second. For velocity, both the direction and magnitude must be specified as it is a vector quantity.

Acceleration is the increase of velocity expressed in feet or centimeters per second per second.

All machines—however complicated—can be reduced to six simple forms:

1. The *lever* consists of a bar free to turn around a point, called the fulcrum.
2. The *wheel and axle* may be considered a rotating or continuous lever; it may consist of a large wheel and a small wheel attached together; or of a wheel attached to an axle; or of a handle attached to an axle.
3. The pulley or block and tackle is also a modified form of lever. In its simplest form, it consists of a disc and of a rope placed in a groove on its circumference. It

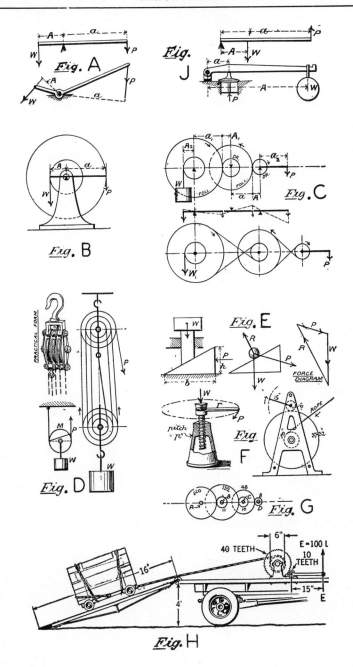

Fig. A

Fig. J

Fig. B

Fig. C

Fig. D

Fig. E

Fig. F

Fig. G

Fig. H

may be considered as a continuously acting lever whose fulcrum is in the middle.

4. The inclined plane is an oblique surface which forms an angle with the base.

two basic machine forms: the lever and the inclined plane.

6. The screw is a spiral or continuous inclined plane.

From this, it is obvious that, in final analysis, there are only two basic machine forms; the lever and the inclined plane.

In any machine, there is a point P where the force is applied and a point W where work is accomplished. Neglecting the resistance, the work done is equal to the applied force.

In a machine, the ratio of the resistance, or load, to the applied force, or effort, is called mechanical advantage. In constructing a machine one of the aims is to obtain the highest possible mechanical advantage.

Figures A and J show various forms of levers. Figure B illustrates a wheel and axle. Figures C and G are gear trains. Figure D shows the pulley, or block and tackle. Figure E illustrates the principles of the inclined plane, and figure F shows a screw combined with a lever.

A compound machine is the combination of simple machines to give greater mechanical advantage. The mechanical advantage of a compound machine is the product of the mechanical advantages of the individual machines that make up the compound machine. Figure H shows a combination of a crank, axle, and inclined plane.

Every machine performs at least one of the following functions:

1. It changes the applied force or effort
2. It changes the direction of the applied force
3. It changes the speed
4. It transmits the force from one opint to another.

There is no machine that could deliver work without work being spent on it. Moreover, the work delivered by a machine is always less than the work supplied, since some work is lost by overcoming resistance, which is usually friction. Nevertheless, machines are powerful tools of human progress and our modern age would not be possible without them.

PART I—STATICS

CLASS I.   FASTENERS

Section 1a.   Marine and Masonry Anchors

A—Mushroom anchor.

B—Trawl or sand anchor; fast stock, double fluke.

C—Trawl or sand anchor; loose folding stock.

D—Grapnel.

E—Navy-type swivelling fluke.

F—Denforth anchor.

G—Northill utility anchor.

H—Fisherman's anchor.

J—Laughlin C.Q.R. plow anchor.

K—Anchor trip hook.

L—Rock anchor for guy or suspension bridge.

M—Concrete, sunk in ground with plate and rod reenforcing.

N—Mooring screws, sunk in ground for buoys.

O—Anchor plate, sunk in ground for attaching tie rods and guys.

P—Wall eye, cast to form brick.

Q—Wall eye, built in.

R—Foundation-bolt head, jagged.

S—Foundation bolt with key.

T—Foundation bolt, standard.

U—Rope-pulley leader anchor, knife-wheel grip in ground.

V, W, X—Fencing posts in ground.

Y—Miscellaneous foundation-bolt anchors.

Z—Expansion bolts.

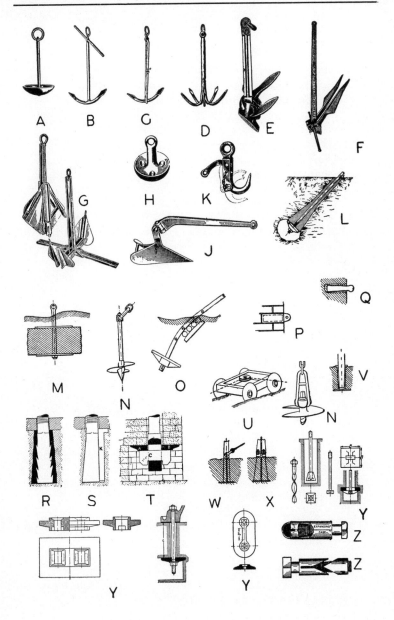

CLASS I.   FASTENERS

Section 1b.   Expansion Shields and Anchors

A—Spring-wing toggle bolt, shown open and closed.

B—Spring-wing toggle bolt: *A* with round head; *B* with square nut; *C* with flat-head screw.

C, D—Spring-wing toggle bolt inserted into a drilled hole in tile or gypsum walls.

E—One-piece toggle, without springs.

F—Paine lead expansion anchors for use in concrete, stone, marble, tile, slate, etc.

G—F in position.

H—Single machine-bolt shield with two-side expansion for use in concrete, etc., installed without a setting tool.

J—Double machine-bolt shield; installed without a setting tool.

K—Fiber or rawhide wood and lag screw anchor for use in brick, plaster, concrete, etc.; the hole need not be plumb; no setting tool required; the fiber anchor should be as long as the threaded part of the wood or lag screw and have the same diameter as the screw.

L, M—Paine steel expansion shells; may be used with two cups; no setting tool required; the hole need not be plumb.

N—Four-point star drill for making expansion-anchor holes in masonry.

O—Paine pipe hook, snug-fit type.

P—Paine adjustable combination pipe hanger; consists of a six-inch length of perforated hanger iron with a gimlet-pointed lag screw at one end and a pipe ring at the other end.

Q—BX staple.

R—Flattened-end lag screw with bolt for use with a malleable expansion shield.

S, T, U—Rawl drive expansion plugs.

V, W—Two-hole and one-hole straps for supporting wall pipe conduit and armored cable.

X—Gimlet-point lag screw; commercial sizes vary in length from $1\frac{1}{2}$ inches to 12 inches and in diameter from $\frac{1}{4}$ inch to $\frac{7}{8}$ inch; lag screws are measured from under the head to the extreme point.

Y—Seebco Scruin patented expansion shield.

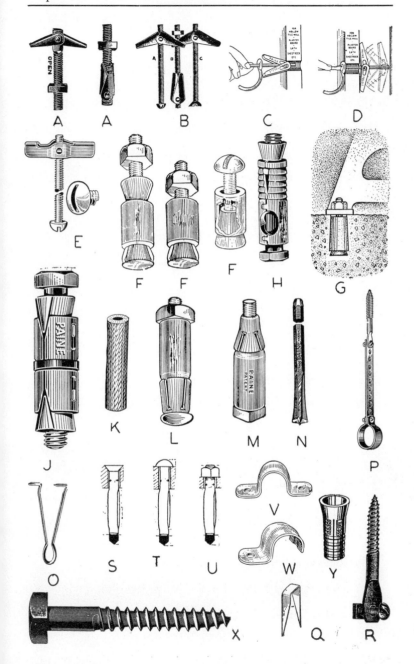

A   A   B   C   D

E   F   F   F   H   G

J   K   L   M   N   P

O   S   T   U   V   W   Y

X   Q   R

CLASS I.   FASTENERS

Section 2a.   Screw and Nut Design Types

Locking Nuts

A, B—Stop plates for locking nuts.

C—Spring-lock washer (Grover patent).

D—Lock nut with pin, square hole and bent plate.

E—Shakeproof lock washer.*

F, G—Pin and groove lock nut, castellated.

Studs

H—Ordinary stud.

J—Stud with square collar.

K—Forcing screw.

L—Adjusting screw.

M—Setscrew with saddle.

N—Double-nutted bolt.

O—Stud and screw.

P, Q, R—Tap and stud bolts.

S, T, U—Through bolt with nuts at both ends.

V—Machine screws.

W—Setscrew with conical point.

X, Y—Setscrew with metal-pad shaft protection.

Z—Miscellaneous screws and bolts.

---

* SHAKEPROOF is a registered trade mark of the Illinois Tool Works.

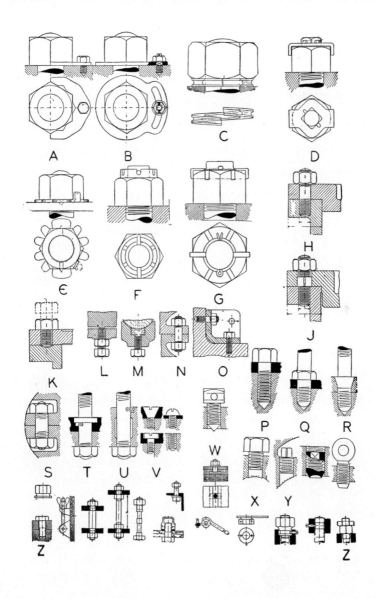

A B C D

E F G H

J

K L M N O P Q R

S T U V W X Y

Z

Z

CLASS I.   FASTENERS

Section 2b.   Screw and Nut Design Types

Threads

A—American National Standard thread.

B—Brown and Sharpe 29° worm thread.

C—Whitworth's thread.

D—Sellers thread.

E—Sharp V thread.

F—Square thread.

G—Buttress thread.

H—Acme standard thread.

J—Round or knuckle thread.

Special Bolt Heads

K—Eye bolt.

L—Hook bolt.

M—Countersunk head.

N—Special head.

O—T-head.

P—Wedge head.

Q—Eye bolt.

R—Boss-head bolt.

S—Lifting-eye bolt.

T—Conical-head bolt.

U—Bolt with intermediate head.

Nuts

V—Hexagonal nut, chamfered at one end.

W—Hexagonal nut, chamfered at both ends.

X—Hexagonal nut, flanged.

Y—Hexagonal nut, spherical at one end

Z, AA—Cap nuts.

BB—Square nut.

CC—Fluted nut.

DD—Circular-back nut.

EE—Fluted nut.

FF—Screw-driver nut.

GG—Capstan nut.

HH—Spring pawl.

JJ—Gravity pawl.

KK—Nut with stop pin.

LL—Wile's lock nut.

MM—Nut with split pin.

NN—Nut with taper pin, split.

OO—Locking plate.

PP—Lock washer.

QQ—Penn or ring nuts.

RR—Nut with setscrew.

SS—Condenser ferrule.

TT—Pin nut.

UU—Nut with holes for a forked spanner.

VV—Stud driver.

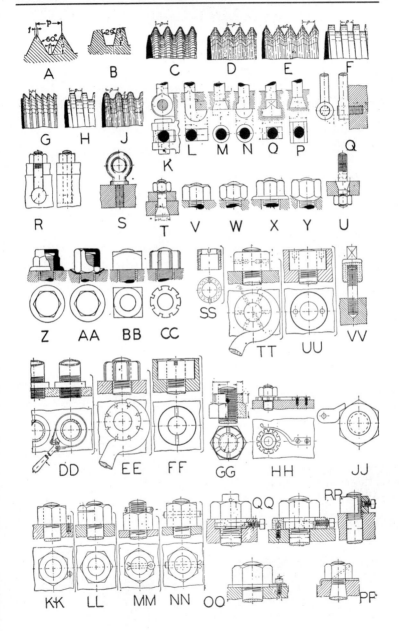

CLASS I.   FASTENERS
Section 2c.   Commercial Nuts

A—"Flexloc" combination-lock and stop nut with National coarse or fine thread.

B—Elastic self-locking stop nut; the standard thickness shown; shaded cross-section shows elastic locking collar which is smaller than the bolt diameter.

C—Stover self-locking nut; the thread is oval to fit over a round bolt thread.

D—Speed nut (Timmerman Products, Inc.) for use on any type of screw or bolt.

E—Crimp nut (Diamond patent); shown inserted and ready.

F—Crimp nut, partly expanded.

G—Crimp nut, completely attached.

H—Watertight-seal grommet for bolts.

J—Hot-pressed square nut, American Standard heavy (U.S.S.).

K—Hot-pressed hexagon nut, American Standard heavy or heavy jam.

L—Hot-pressed square nut, American Standard regular.

M—Hot pressed hexagon nut, American Standard regular.

N—Cold-punched square nut, American Standard heavy.

O—Cold-punched hexagon jam nut, American Standard heavy jam (U.S.S.).

P—Cold-punched hexagon nut, American Standard heavy (U.S.S.).

Q—Cold-punched square nut, American Standard regular.

R—Cold-punched hexagon jam nut, American Standard.

S—Cold-punched hexagon nut, American Standard regular.

T—Semifinished heavy hexagon nut, single chamfer, single countersink; available in all sizes.

U—Semifinished regular hexagon nut.

V—Semifinished light hexagon nut.

W—Wing nut.

X—S.A.E. Steel lock washer.

Y—Cotter pin with extended prong.

Z—Dock washer.

AA—Square washer, wrought iron.

BB—Turnbuckle with right- and left-hand threads.

CC—Clevis, available with right- or left-hand thread, American Standard coarse-thread series, class 2.

DD—Gripper lock nut used in Great Britain.

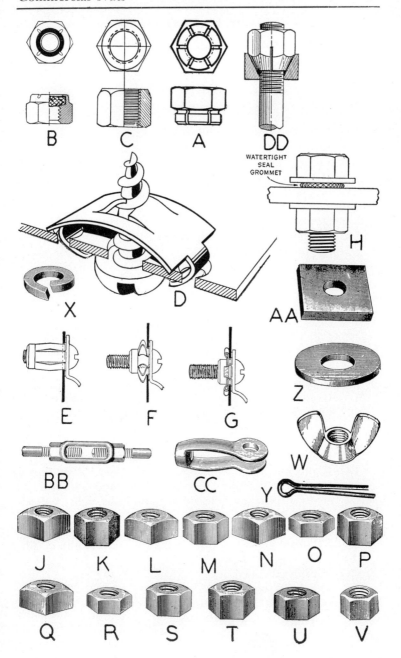

B C A DD

WATERTIGHT
SEAL
GROMMET

H

X D AA

E F G Z

BB CC W

Y

J K L M N O P

Q R S T U V

CLASS I. FASTENERS
Section 2d.   Miscellaneous Bolts and Screws

A—Bolt with a head requiring special spanner or pointed bar.

B—Cylinder-head bolt with drilled holes and special spanner.

C—Cylinder-head bolt with flutes for the spanner.

D—Cylinder-head bolt with two flat surfaces to fit the standard spanner wrench.

E—Socket-head bolt for receiving a screw.

F—Milled-head screw.

G—Bolt with a head for a forked spanner.

H—T-head bolt.

J—Hexagon-collar bolt.

K—Hexagon-head bolt with collar.

L—Eye bolt with flat sides.

M—Hook bolt.

N—Lewis bolt for concrete.

O—Rag bolt.

P—Cottered bolt.

Q, R, S—Lewis bolts and key pieces.

T—Collar stud.

U—Split-spring-head bolt.

V—Hook bolt.

W—Solid-head and collar bolt (bed bolt).

X, Y—Heads for bolts to slide and turn in T-grooves of planing machines.

Z—Countersunk bed bolt (boiler stay).

AA—Ring coupling.

BB—Right- and left-hand screw couplings for tie rods.

CC—Right- and left-hand screw couplings with halved ends to prevent turning; they may have one fine and one coarse thread for differential motion, or right- and left-hand threads.

DD—Belt screw.

EE—Ball-head bolt and nut; it may be drawn out of line.

FF—Universal bolt head.

GG—Flush-head coned bolt.

HH—Mutilated screw and nut.

JJ—Coned bolt for securing and keying two parts of a machine in exact relation.

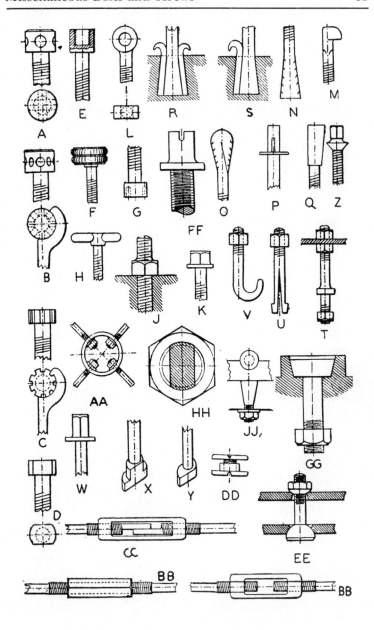

CLASS I.   FASTENERS

Section 2e.   Miscellaneous Screws, Bolts and Nuts

A—Compound nuts and lever for great leverage on a screw, as in a press; one nut arm is used as a fulcrum by which the lever forces the other nut around; stepped pawls prevent the first nut from being loosened while moving the second nut.

B—Screw-eye and handle nut.

C—Sunk setscrew with differential threads for drawing plates together.

D—Mutilated screw to slide into a nut having corresponding sections of the thread cut away; to be set by a partial turn; used in breech pieces of artillery.

E—Bolt head with transverse holes.

F—Backlash nut for square-thread screw.

G—Slotted nut and set pin for fine adjustment, or for taking up wear.

H—Split nut with bolt to clamp it.

J—Bolt with triangular washer to prevent loosening of the nut.

K—Tapered-thread right- and left-hand coupling.

L—Intermittent worm; part of it is without pitch so as to give a pause to the wheel it meshes with.

M—Felloe wood screw.

N—Oval fillister-head wood screw.

O—Countersunk fillister-head wood screw.

P—Square-bung-head wood screw.

Q—Winged wood screw.

R—Headless wood screw.

S—Round-bung-head wood screw.

T—Dowel.

U—Close-head wood screw.

V—Cone-point wood screw.

W—Gimlet-point wood screw.

X—Diamond-point wood screw.

Y—Pinched-head wood screw.

Z—Drive wood screw.

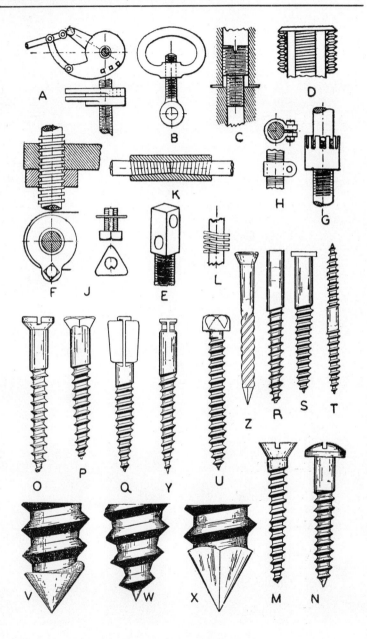

CLASS I.   FASTENERS

Section 2f.   Commercial Bolts (Bethlehem Steel Co.)

A, B, C—Eye bolts.

D, E, F, G—Hook bolts.

H, J—U-bolts.

K—Oval T-head superheater bolt.

L—Upset forging.

M—Guard-rail anchor rod.

N—Insulator pin.

O—Gland nut.

P—Pole step.

Q—Liner bolt.

R—Gate hook.

S—Silo rod.

T—Double-end rod.

U—Upset rod.

V—Pipe band.

W—Hook-head bolt.

X—Collar bolt.

Y—Cap nut.

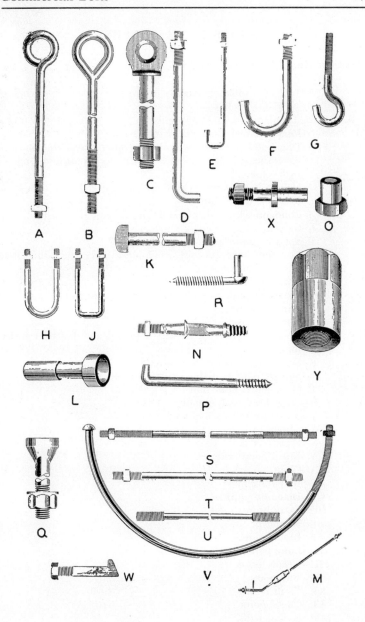

CLASS I.   FASTENERS

Section 2g.   Nails and Spikes

A–Railroad spike; available in sizes of $\frac{5}{16} \times 2$ inches to $\frac{5}{8} \times 6\frac{1}{2}$ inches.

B, C, D, E–Tie-plate screws (screw spikes); $\frac{3}{4}$ inch diameter, 5 inches long to 1 inch diameter, 7 inches long.

F, G–Diamond- and button-head boat spikes; available in sizes of 5 inches long, $\frac{5}{16}$ inch square to 16 inches long, $\frac{3}{4}$ inch square.

H–Cut nail; available in sizes 2d (pennyweight), 1 inch long, to 60d, 6 inches long.

J–Common wire nail; available in the same sizes as H.

K–Carpet tack.

L–Sprig or dowel pin.

M–Common brad (casing); available in sizes 2d, 1 inch long to 60d, 6 inches long.

N–Common brad (finishing).

O–Flooring brad; available in sizes 6d, 2 inches long to 20d, 4 inches long.

P–Common brad.

Q–Boat nail; available in sizes 4 to 20d and $1\frac{1}{2}$ to 4 inches long.

R–Slating nail; available in sizes 2 to 5d.

S–Shingle nail; available in sizes 3 to 4d.

T–Hook-metal lathe nail.

U–Large-head roofing nail.

V–American felt roofing nail.

W–Clinch nail.

X–Cigar-box nail.

Y–Blued hoop fastener.

AA–Clout.

BB–Berry-box nail.

CC–Diamond point.

DD–Extra blunt point.

EE–Round point.

FF–Blunt point.

GG–Long point.

HH–Needle point.

JJ–Side point.

KK–Clout point.

LL–Chisel point.

MM–Special nails.

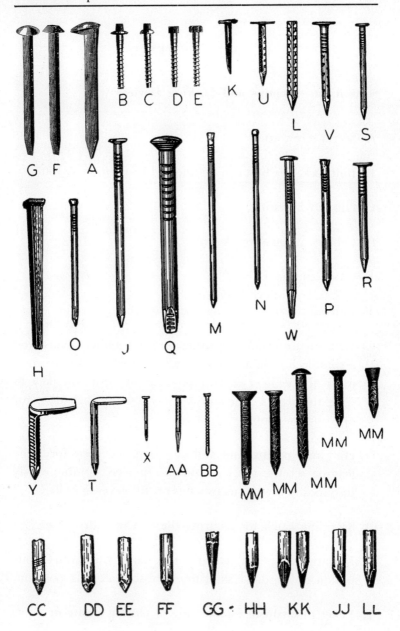

G F A B C D E K U L V S

H O J Q M N W P R

Y T X AA BB MM MM MM MM MM

CC DD EE FF GG HH KK JJ LL

CLASS I.   FASTENERS

Section 3a.   Clamps and Locking Devices

A—Secret screw attachment.

B—Bolt lock; the bolt can be released by turning at 180°.

C—Gib key fastening.

D—Gib key using a wooden bar.

E—Catch and hook.

F—Hinged catch for locking a screwed gland or nut.

G—Combination lock consisting of any number of discs on a spindle having a feather key, arranged so that the discs must all be in a certain position to allow the key to slide through a notch or keyway, cut in each disc, for opening the loop.

H—Half-nut locking and unlocking device used for lathe leading screws; the half nuts are moved simultaneously in opposite directions by cams on the lever spindle.

J—Swinging catch for securing the end of a drop bar.

K—Locking screw for locking the hand wheel and spur pinion to the drive shaft to establish operating position.

L—Lever mechanism for opening or closing a gate or door.

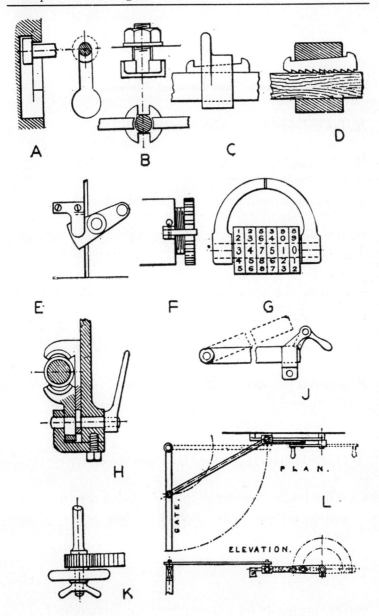

A

B

C

D

E

F

G

J

H

K

GATE.

PLAN.

ELEVATION.

L

CLASS I.   FASTENERS

Section 3b.   Clamps and Locking Devices

A—Locked shutter bar with swinging pawl.

B—Drop catch for a bar.

C—Wedge-plate and screw fastening for cutters.

D—Hinged bolt and handle nut for locking and tightening a door.

E—Handle nut with tapered flange for fastening down flat work.

F—Parallel bar movement for railway switches.

G—Setscrew fastening for flange and socket.

H—Locked center pin.

J—Hinged handle, combined latch and staple.

K—Wire-ring locking device, the wire being driven into the groove through holes drilled in line with the wire groove.

L—Cotter for locking a sliding spindle.

M—Locking device for a spring lever, e.g., button hook.

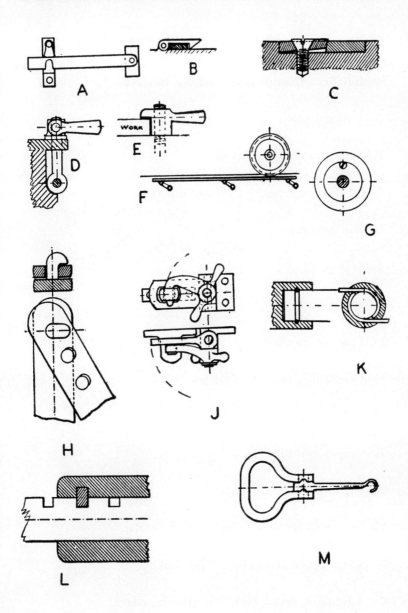

CLASS I.   FASTENERS

Section 3c.   Clamps and Locking Devices

A—Locking pawl.

B—Spring catch for swing doors.

C—Spring catch, beveled on one side only

D—Disc and pin.

E—Side pawl.

F—Common latch.

G—Cam locking bolt.

H—Crank-movement locking bolt.

J—Bolt of common lock.

K—Common sliding bolt.

L—Drilling-machine clamp.

M—Automatic bench clamp for work on flat.

N—Screw bench clamp for cabinetmakers.

O—Automatic bench clamp for planing edges.

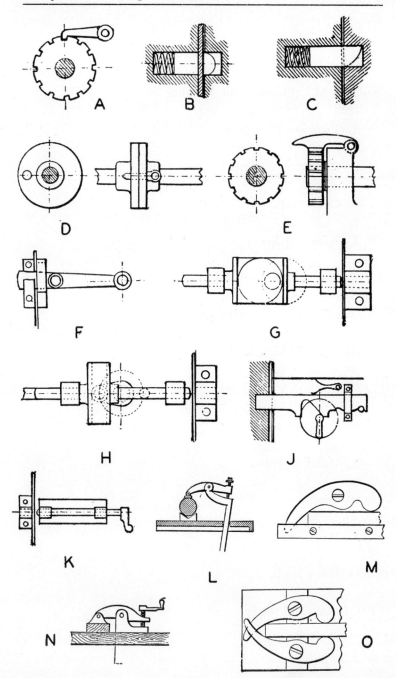

CLASS I.   FASTENERS

Section 3d.   Clamps and Locking Devices

A—Twisting a flat bolt.

B—Rod or rope stopper with cam-lever grip.

C—Chain stop.

D, E—Spindle grips.

F—Clamp and screw.

G—Sliding-shaft locking pin.

H—Lever-locking hook.

J—Bow catch for ladles.

K—Cross bar and hooks.

L—Hand setscrew.

M—T-catch.

N—Hasp and staple.

O—Hook latch.

P, Q—Hasp and staple.

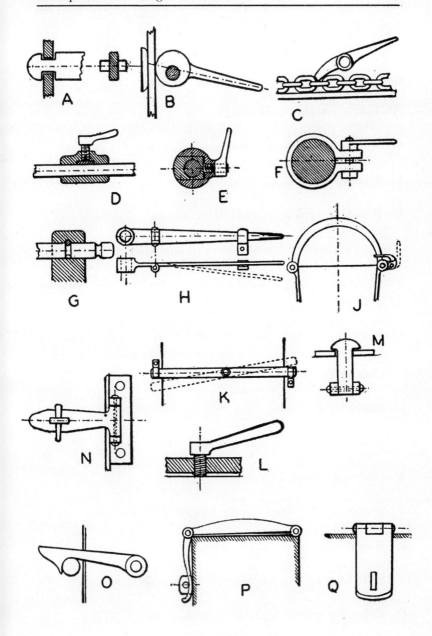

CLASS I.   FASTENERS

Section 3e.   Clamps and Locking Devices

A—Trap-door automatic catch.

B—Screw and bridle suspension.

C—Spring-stud lock.

D—Radial hinged-lever and crown ratchet.

E—Locking bar for fixing a lever in any position.

F—Crank-arm device for locking a valve or lever in two
    positions.

G—Pawl for locking sliding shafts; it is used for winches
    and has a double and a single purchase gear or shifting
    clutches.

H—Fastening eye bolt.

J—Door-fastening staple or cotter.

K—Common cotter.

L—Roller and inclined slot for locking a rope or rod.

M—Revolving-bush lock.

N—Wire-fencing notches in angle or channel iron.

O—Releasing grip of a pile-driving machine.

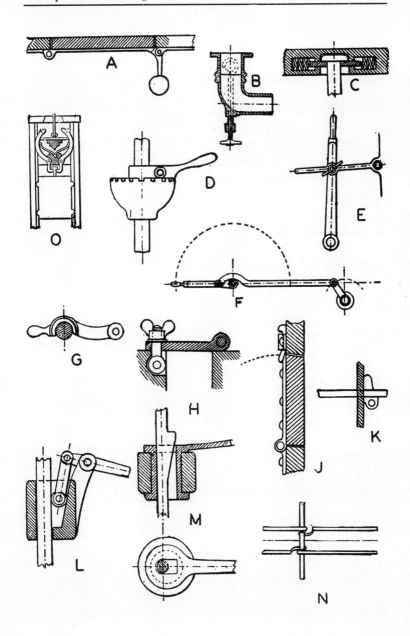

CLASS I.   FASTENERS

Section 3f.   Clamps and Locking Devices

A—Radius bar.

B—Horseshoe distance piece to be placed between a sliding pinion and a shaft collar to keep them in or out of gear.

C—Coned screw lock.

D—Setscrew fitting.

E—Split block for gripping a rod.

F—Cam catch for locking a wheel or spindle.

G—Locking gear for a shaft driven by spur gearing, used in place of a clutch.

H—Locked nut.

J—Spring-pawl umbrella catch.

K—Spring snap.

L—Locking pawl for spur teeth.

M—Spring spindle.

N—Spring pawl that locks a wheel against moderate force, but gives way to a greater force.

O—Locking device for a lathe headstock or tool rest.

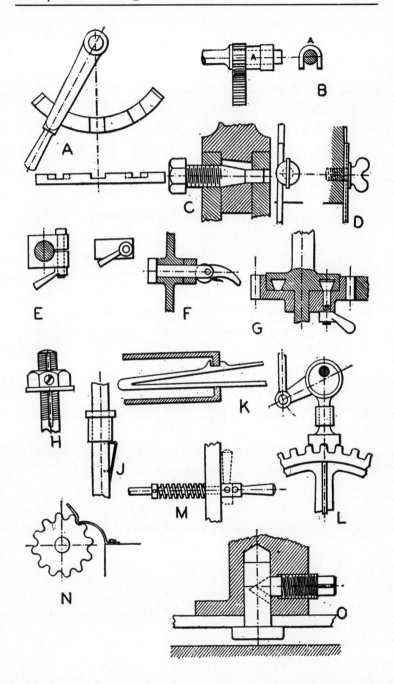

CLASS I.   FASTENERS

Section 4a.   Holders and Clamps

A—Elevating tool box.

B—Split-bar grip, or tool holder.

C—Hand pad for holding small tools.

D—Revolving tool post, or capstan head.

E—Sockets for various tools with parallel or tapered holes; for parallel holes a key or set-screw is used.

F—Adjustable tap wrench.

G—Lathe carrier, for round rods and spindles.

H—Tool box, with two stool stocks and setscrews sliding in T-grooves in the slide rest.

J—Tool box with clamping screw and plate, which can be turned at any angle.

K—A modification of J.

L, M, N, O, P—Clamping devices for jigs.

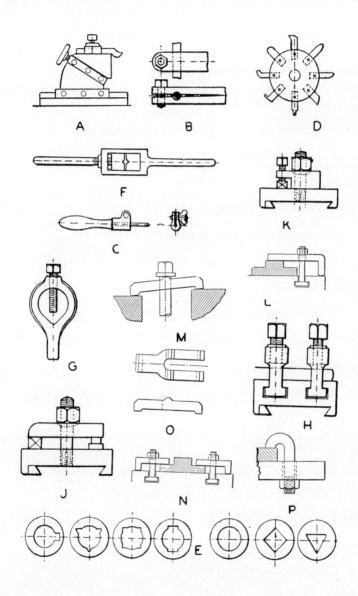

CLASS I.   FASTENERS

Section 4b.   Holders and Clamps

A—Adjustable-grip holder for bricks, etc.

B—Grapnel with shear pins for excessive strains.

C—Parallel-vice grip.

D—Square-hole central grip.

E—Tool-holder-gripping device.

F—Spring grip for small drills, pencils, pins, etc.

G—Split tool bar with transverse cutter.

H—Socket and setscrew for drills.

J—Split tool holder (Barber's patent).

K—Simplest form of V-grip.

L—Screw clamp.

M—Rod clip.

N—Screw clamp.

O—Adjustable tool box.

P—Drill socket with diagonal pin for gripping the shank of
the drill.

Q—Toothed V-grip for chucks.

R—Cam-clamp for jigs.

S--Gripping-dog clamp for jigs.

T—Simplest form of V-grip.

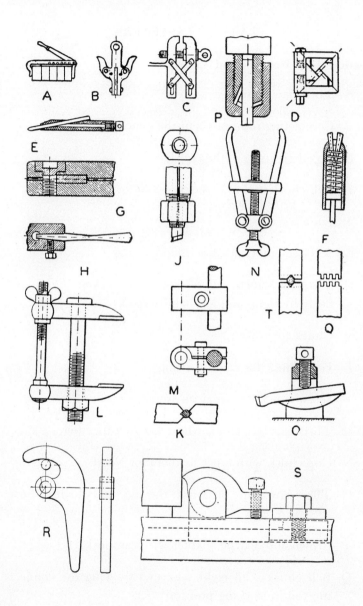

A

B

C

P

D

E

G

J

N

F

H

T

Q

L

M

K

O

R

S

CLASS I.   FASTENERS

Section 4c.   Holders and Chucks

A—Three-jaw chuck.

B—Connecting-rod brasses in the chuck.

C, D—Eccentric strap in the chuck.

E—Bell chuck and four setscrews for a lathe.

F—Four-jaw chuck.

G—Screwed-cup chuck for wood turning.

H—Two-jaw chuck whose jaws are moved simultaneously by right- and left-hand screws.

J—Chuck for wheels.

K—Fork chuck for wood turning.

L—Screw chuck for wood turning.

M—Three-jaw-face chuck whose jaws act together.

N—Cup chuck with taper feathers for wood.

O—Three-jaw guide or chuck, the slide jaws being adjustable by screws.

P—Brown and Sharpe adapter with cam lock.

Q—Bell chuck with eight setscrews showing the connecting rod in working position.

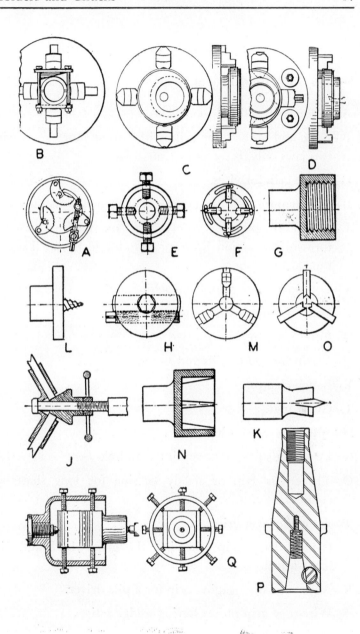

CLASS I.   FASTENERS

Section 4d.   Holders and Grips

A—Collar grip and bolt or setscrew.

B—Cone and screw lever grip, with two or more jaws; with two jaws only, it serves as a small vice.

C—Taper grip for vices.

D—Rail grip for holding a crane or car down to its railway.

E—Cam-lever grip for safety gear on inclines; usually thrown into action by a spring released by breakage of the hauling rope.

F—Cone-centering grips for machine tools.

G—Hinged clamp with screw and nut.

H—Fitter's clamp.

J—V-grip for round rods and tubes.

K—Bench clamp.

L—Grip tongs.

M—Split-cone expanding chuck.

N—Le Count's patent expanding mandrel.

O—Three-jaws grip or steady bearing for long shafts or spindles.

P—Eccentric lever grip.

Q—Cable-railway grip; showing the grip wheel and the hand wheel.

R—Automatic disengaging grip for a pile driver.

S—Wire-rope grip pulley having wedge action.

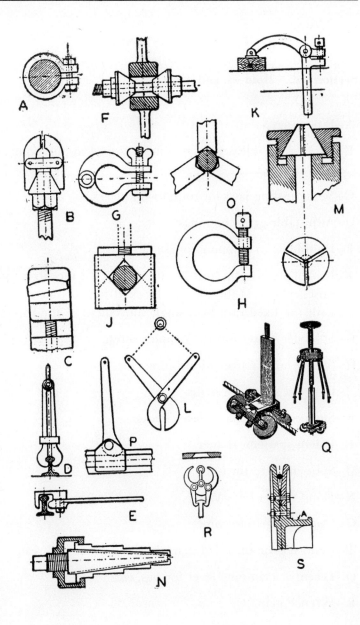

CLASS I.   FASTENERS

Section 4e.   Holders and Grips

A—Pipe tongs.

B—Paper grip, released by striking a stop A at any point of
its travel.

C—Self-adjusting jaws for round workpieces.

D—Adjustable gripping tongs.

E—Double-screw gripping tongs.

F—Instantaneous grip for vice; the worm A is eccentric, and
raises or lowers the toothed block B into or out of gear
with the fixed rack by a single movement of the handle.

G—Split sleeve and nut for gripping a rod.

H—Spring taper socket with sliding ring.

J—Cap and socket for drills.

K—Stepped jaw for lathe-face chucks.

L—Patent spanner (Bauer).

M—Split-end grip for rods.

N—Hand screws for a V-grip.

O—Double V-grip for pipes.

P—Central grip, thread A being one half pitch of thread B.

Q—Eccentric clamping lever for jigs, etc.

R—Wrench grip.

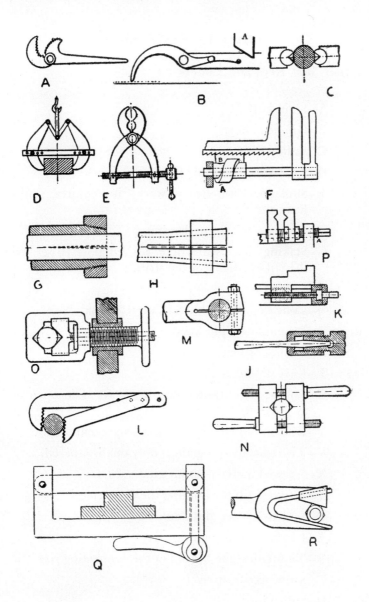

CLASS I.   FASTENERS

Section 5a.   Keys and Cotters

A—Plain key.

B—Gib-head key.

C—Saddle or hollow key for light work only.

D—Key on flat, suitable for heavier work.

E—Sunk key for heavy work; it requires skillful fitting.

F—Key boss for strengthening.

G—Staking on, used where a solid boss passes over the enlarged end of the shaft.

H—Staking on.

J—Cone keys for light work on frictional drives.

K, L, M—Round-taper pin keys.

N—Sliding or feather key.

O to T—Keys similar to N.

U—Cotter; *A, C* causes strap *E, F, G, H* to open out as per dotted line.

V—Gib combined with a cotter to prevent strain.

W—Two gibs used together; only one is tapered.

X—Screwed cotter without gib.

Y—Screwed gib for holding the cotter.

Z—Screwed cotter and double gibs.

AA—Use of setscrew with cotter.

BB—Tightening the brasses of the connecting rod.

CC—Screw-fixed cotter.

DD—Split pins.

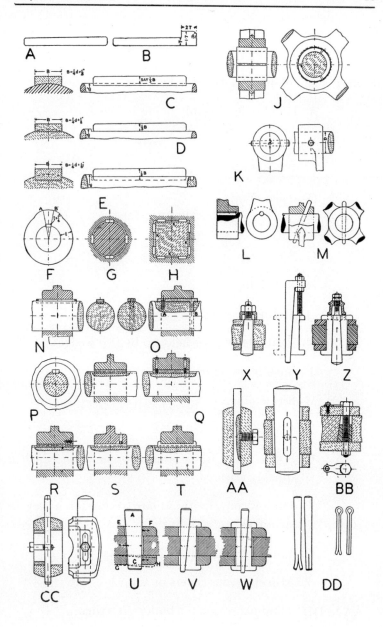

CLASS I.   FASTENERS

Section 6a.   Manholes, Hand-holes, Covers, Doors

A—Hand-hole assembly being inserted.

B—A in place.

C—Wrought-iron or steel manhole door, dished.

D—Oven door.

E, F—Soot door.

G—Screw cap or cover.

H—Revolving door.

J—Sliding door for furnaces.

K—Manhole or door held up by two wedges.

L—Steel-plate lid or cover for tanks.

M—Screw plug.

N—Manhole door.

O—Screw fixing for a plug, door or valve which is quickly released.

P—Hollow plug with square recess for a key or spanner.

Q—Funnel plug for filling oil tanks.

R to Z—Miscellaneous covers or lids.

AA to DD—Manhole covers for boilers (German).

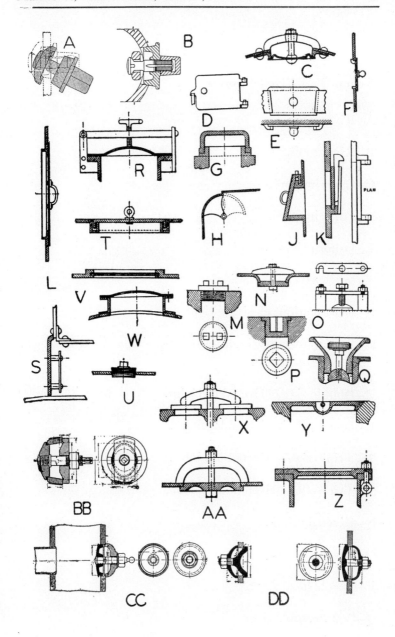

CLASS I.   FASTENERS

Section 7a.   Riveted Joints

A—Rivet types.

B—Rivet tools.

C—Caulking tool.

D—Fullering tool.

E—Osborn rivet symbols.

F—Single-riveted lap joint.

G—Deformation of single-riveted joint, caused by tension.

H—Butt joint with single butt strap single riveted, showing also deformation.

J—Double zig-zag-riveted lap joint.

K—Double chain-riveted lap joint.

L—Triple zig-zag-riveted lap joint.

M—Triple chain-riveted lap joint.

N—Riveted crimp.

O—Riveted scarf.

P—Steeple-head rivet.

Q—Countersunk rivet.

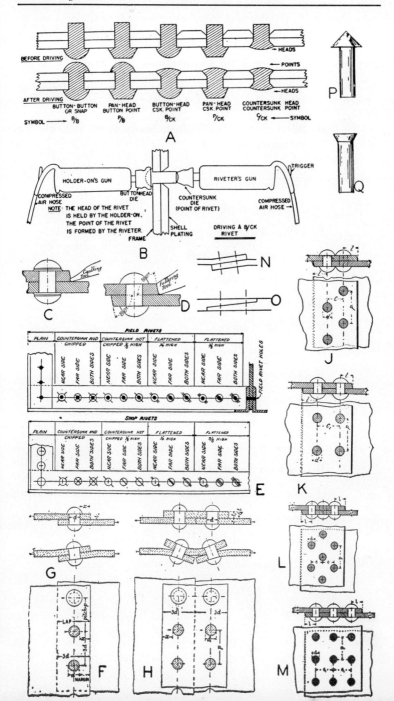

CLASS I.   FASTENERS

Section 7b.   Boiler Flue Connections and Stays

A—T-ring, very rigid longitudinally; rivet heads are exposed to fire.

B—Bowling ring or Bolton hoop, more flexible longitudinally.

C—Adamson's ring, flexible longitudinally, rigid circumferentially.

D—Flue-stiffening ring.

E—Davey-Paxman joint.

F, G—Flue-stiffening rings.

H, J, K, L—Fire-box connections and parallel plates.

M—Gusset stays.

N—Diagonal stays.

O—Crowfoot diagonal stays.

P—Through stay.

Q—Solid stay, riveted.

R—Hollow stay, riveted.

S—Flexible stay (Flannery).

T—Various locomotive fire-box stays.

U—Welded butt V-joint used to prevent buckling.

V—Inserting a tap rivet.

W—Reaming out misaligned rivet holes.

X—Plugging and redrilling a rivet hole.

Y—Girder stay.

Z—Dog stay.

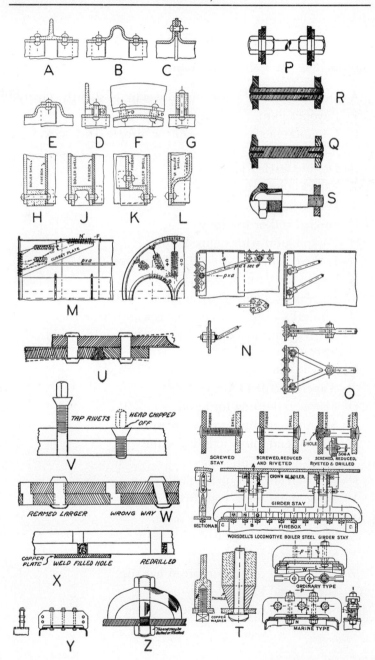

CLASS I.   FASTENERS

Section 8a.   Miscellaneous Joints

A—Dovetail metal joint; the dovetail is slightly tapered.

B—Bayonet joint.

C—Common male and female or nipple and socket rod joint.

D—Double-scarfed joint.

E—Knuckle joint for levers.

F—Universal joint.

G—Scarfed rod or bar joint.

H—Forked joint with stepped gibs.

J—Forked joint and swivel block for screw attachment.

K—Swivel joint for pipe work.

L—Screw socket and spigot joint for rods.

M—Conical socket joint and setscrew.

N—Scarfed joint locked by a cross cotter and tapered ferrule.

O—Swivel joint.

P—Tapered drill socket.

Q—Felton's coupling for a rod.

R—Sheet-metal joint, grooved seam.

S—Sheet-metal joint, standing seam.

T—Sheet-metal joint, slip.

U—Sheet-metal joint, drive slip.

V—Sheet-metal joint, end slip.

W—Sheet-metal joint, double seam.

X—Sheet-metal joint, Pittsburgh seam.

Y—Sheet-metal joint, bar slip.

Z—Sheet-metal joint, reinforced bar slip.

AA—Sheet-metal joint, pocket slip.

BB—Sheet-metal joint, angle connection.

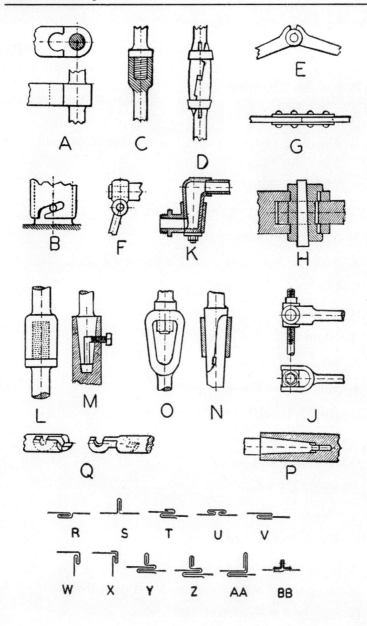

A

C

D

E

G

B

F

K

H

L

M

O

N

J

Q

P

R   S   T   U   V

W   X   Y   Z   AA   BB

CLASS I.   FASTENERS

Section 9a.   Welding

A—Welding circuit (Lincoln Electric Co.).

B—Shielding of the arc and slag protection of weld metal while cooling.

C—Electric welding.

D—Examples of welds and locations.

E—Square butt joint.

F—Single-U butt joint.

G—Double-U butt joint.

H—Single-fillet lap joint.

J—Double-fillet lap joint.

K—Double-J T-joint.

L—Single-bevel T-joint.

M—Double-bevel T-joint.

N—Single-V butt joint.

O—Double-V butt joint.

P—Single-J T-joint.

Q—Square T-joint.

R—Finish corner joint.

S—Half-open corner joint.

T—Oxygen and acetylene torch for cutting steel.

U—Staggered intermittent fillet welds.

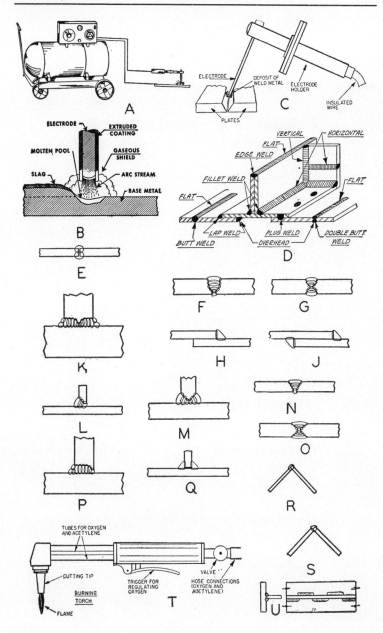

A

C

ELECTRODE — DEPOSIT OF WELD METAL — ELECTRODE HOLDER — INSULATED WIRE — PLATES

B

ELECTRODE — EXTRUDED COATING — MOLTEN POOL — GASEOUS SHIELD — SLAG — ARC STREAM — BASE METAL

D

VERTICAL — HORIZONTAL — FLAT — EDGE WELD — FILLET WELD — FLAT — FLAT — LAP WELD — PLUG WELD — DOUBLE BUTT WELD — BUTT WELD — OVERHEAD

E

F

G

K

H

J

L

M

N

P

Q

O

R

S

T

TUBES FOR OXYGEN AND ACETYLENE — CUTTING TIP — BURNING TORCH — FLAME — TRIGGER FOR REGULATING OXYGEN — VALVE — HOSE CONNECTIONS (OXYGEN AND ACETYLENE)

U

CLASS I.   FASTENERS

Section 9b.   Fabricated Welded Shapes

A—Standard-shape channel bar.

B—T-bar.

C—Angle bar.

D—I-beam.

E—H-bar.

F—Bulb-angle bar.

G—Z-bar.

H—Flat bar.

J—Bulb T-bar.

K—Special welded fabricated shape, two angles.

L—Special welded fabricated shape, one and two channels.

M—Two T's and one plate.

N—Two channels and two plates.

O—Two channels.

P—Fabricated welded bearings (Lincoln Electric Co.).

Q—Bearing (Lincoln Electric Co.).

R—Crank.

S—Base plate.

T, U, V, W—Levers.

X, Y, Z, AA—Clevises.

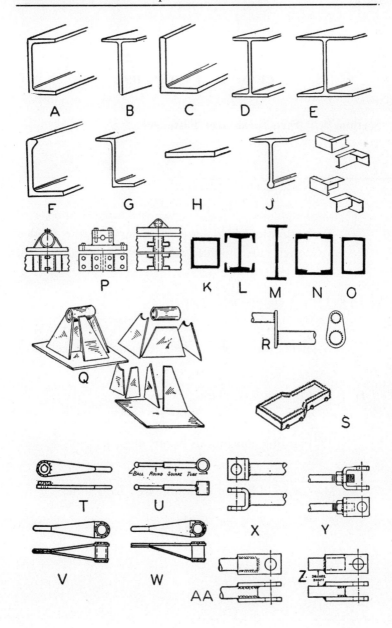

A B C D E

F G H J

P K L M N O

Q

R

S

T U X Y

V W Z

AA

CLASS I.　　FASTENERS

Section 9c.　　Structural-Steel Fasteners

A—Node of riveted truss.

B—Node of welded truss, showing how the welded
design cuts the dead weight as compared with A.

C, D, E, F—Miscellaneous tie rods.

G—Sleeve nuts for tie rods (Bethlehem Steel Co.).

H, J, K—Column base plates.

L—Framing continuous beams to girders.

M—Arc-weld design of double-web bar frame.

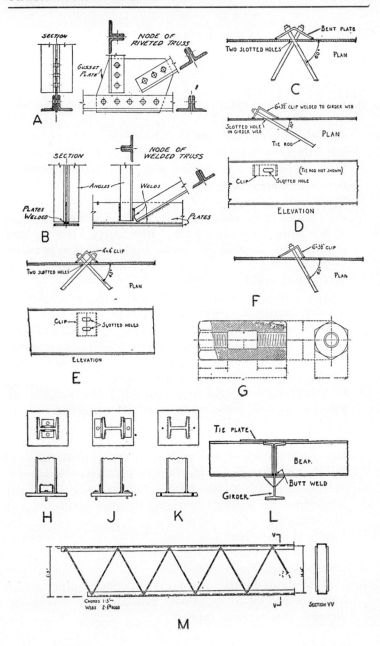

CLASS I.   FASTENERS

**Section 10a.   Screw and Welded Pipe Joints
(Walworth Co.)**

A—Steel or wrought-iron pipe joint with screwed coupling;
available in two types: 1. a straight coupling with taper
thread, necessitating rotation of one pipe, 2. a right and
left coupling without the necessity of rotating either
pipe; limited to the smaller pipe sizes for low-pressure
and low-temperature work.

B—Steel or iron pipe joint with screwed union; neither
pipe needs to be rotated; connections are made in malleable iron, brass, and forged steel; sizes 4 inches and
smaller are suitable for both high- and low-pressure
work.

C—V-type butt-weld pipe joint.

D—V-type butt-weld pipe joint with backing or chill ring
for eliminating icicles or welding pellets; the pipe ends
are beveled, usually at 45°, thoroughly cleaned, and then
brought together until they nearly touch; larger pipes
are tack-welded prior to making complete weld; this
tacking is omitted on the smaller sizes.

E—Cup-weld steel or iron pipe joint; it is particularly ap-
plicable to field welding as it is self-centering and no
clamps are needed for holding the pipe in place as with
regular butt welds.

F—Sleeve-weld steel or iron pipe joint for unusual stresses,
in other variations, in addition to the two welds shown
in the illustration, the reinforcing sleeve is split and
longitudinally welded; or there are spot welds around
the center line of the sleeve.

G—Cast-iron pipe expansion joint.

H—Cast-iron pipe oversized threaded joint.

J—Cast-iron pipe-size threaded joint.

K—Walseal brazed joint for brass or copper pipe; white
indicates special alloy, crosses indicate flux, arrows indicate where heat is applied.

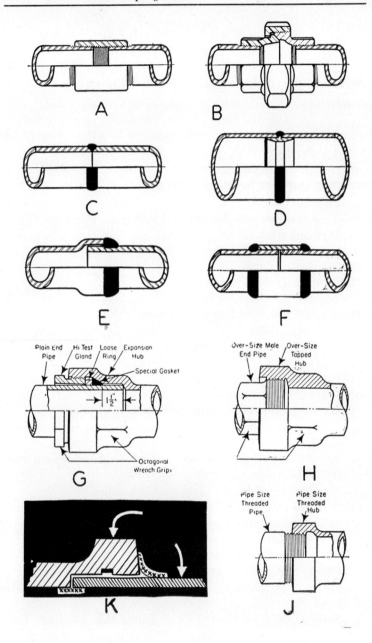

A

B

C

D

E

F

G

Plain End Pipe
Hi Test Gland
Loose Ring
Expansion Hub
Special Gasket
$1\frac{1}{2}"$
Octagonal Wrench Grips

H

Over-Size Male End Pipe
Over-Size Tapped Hub

K

J

Pipe Size Threaded Pipe
Pipe Size Threaded Hub

CLASS I.   FASTENERS

Section 10b.   Fabrication of Welded Piping

A—Design of inside-corner welds on turns; joints miter-cut, wall thicknesses 3/16 inch and over beveled (The Linde Air Products Co.).

B—Design of outside-corner welds; wall thicknesses 3/16 inch and over beveled; joints miter-cut.

C—Full fillet weld.

D—Tack-weld section for beveled joints.

E—Welding neck flange.

F—Van Stone flange with welding nipple.

G—Fabricated flange.

H—Welded screwed flange.

J—Welded standard flange.

K—Welded standard flange with long hub.

L—Welding elbow.

M—Welding elbow with tangents.

N—Multipiece turn 90°; miter joints.

O—Concentric welding reducer.

P—Eccentric welding reducer.

Q—Welding end closure; for highest scale of pressures.

R—Orange-peel end closure; for next highest scale of pressures.

S—Square end closure; lowest on the scale for resisting internal pressure.

T—Header assembly; preparation for welding.

U—Reverse straining to control distortion of large headers.

V—Self-ejecting pipe dolly; used for rotation welding of 200 feet lined-up sections of piping; in level country, 500-foot sections are sometimes lined up.

W—Clamp for joint alignment, the pipe ends to be joined are held in place by the clamp for tack-welding operation.

X—Adjustable pipe clamp (The Linde Air Products Co.).

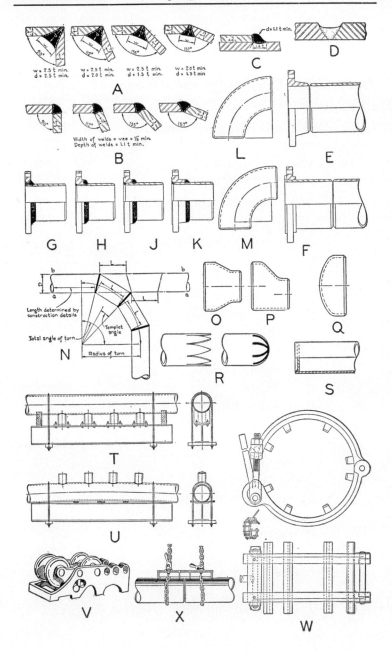

CLASS I. FASTENERS

Section 10c. Pipe Clamps and Hangers

A—Wrought-iron clevis hanger.

B, C—Wrought-iron riser clamp.

D—Blake malleable-iron pipe hanger.

E—Iron extension pipe hanger.

F—Malleable-iron beam clamp.

G—Grinnell constant-support hanger; piping in low-est position.

H—G in highest position.

J—Forged pipe hanger.

K, L, M—Rollers eliminating strains due to expansion and contraction.

N—Trench or ground roller support.

O—Pedestal roller support.

P—Extension hanger.

Q—Trench or ground roller (German design).

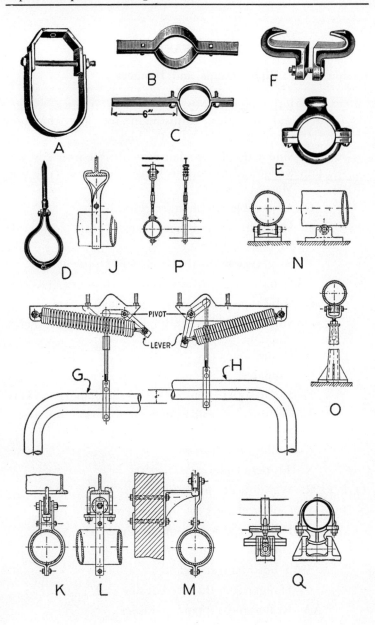

CLASS I.   FASTENERS

Section 10d.   Pipe Clamps and Hangers

A—Full-flange clamp; malleable-iron jaws, wedges and spreader; adjustment for any thickness flange up to 1 inch (Grabler).

B—Side-flange beam clamp; malleable-iron jaw and wedge; adjustable to flange thickness (Fee and Mason).

C—Adjustable ring hanger (Grabler).

D—Malleable-iron hinge hanger with adjustable swivel.

E—Solid-type malleable-iron swivel pipe hanger.

F—Hinged-type adjustable swivel pipe hanger.

G—Solid-type pipe hanger with clevis swivel.

H—Single-roller hanger for pipes of 2 to 20 inches diameter.

J—Single-roller hanger for pipes of 1 to 1½ inches diameter.

K—Single-roller hanger with pipe-hanger sockets.

L—Ceiling-coil roller hanger for two to twelve pipes.

M—Roller on skid base.

N—Trapeze roller.

O, P, Q, R—Wall-bracket suspension.

S, T, U, V—Ceiling flanges for rod suspension.

W—Hanger rod with gimlet point.

X—Hanger rod, threaded at both ends.

Y—T-head hanger rod.

Z—Hanger rod, threaded at one end.

AA—Roller and support (Grabler).

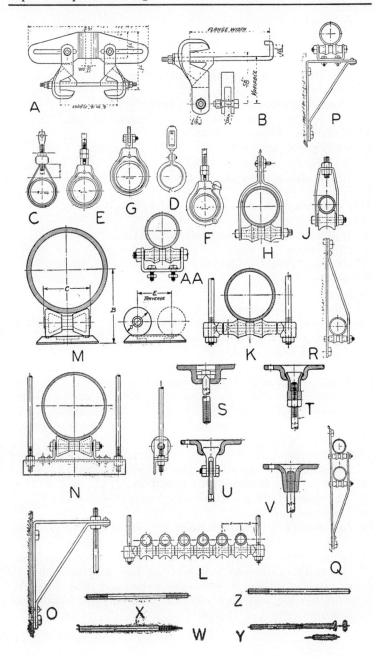

CLASS I.   FASTENERS

Section 10e.   Miscellaneous Pipe Joints

A—Flanged pipe-bend expansion joint.

B—Corrugated-bend expansion joint (German design).

C—Stuffing-box joint, straightaway.

D—Stuffing-box joint, elbow.

E—Stuffing-box joint, straightaway.

F—Single-offset U-bend (Crane Co.).

G—Crossover bend.

H—Expansion U-bend.

J—Double-offset U-bend.

K—Double-offset expansion bend (Crane Co.).

L—Circle bend.

M—90° bend (Walworth).

N—180° bend.

O—Angle bend.

P—Offset bend (Walworth).

Q—Flexible pipe joint; the internal surface of the hub is spherical; corrugated pipe is inserted, and the space is filled with molten lead and caulked.

R—Flexible pipe joint, ball and socket type.

S—Toggle-clip hose coupling or pipe joint.

T—Flexible ball joint; the flanges are spherical and the space is packed with lead ring gasket.

U—Flexible pipe joint; the lead joint consists of a lead gasket.

V—Expanding pipe stopper.

W—Titeflex-metal flexible hose with male pipe fitting at one end and S.A.E. flared-type union at the other end.

X—Titeflex-metal flexible hose application.

Y—Magnilastic 1,000 psi stainless-steel expansion joint.

Z—Universal pipe joint.

ZZ—Installation of condenser tubes; by rolling at one end packing at the other end with metallic packing, without ferrules. Other methods of installation are rolling at one end, packing at the other end with fiber or lace packing and with ferrules or rolling at both ends with fixed tube sheets, or with flexible joint between the shell and one tube sheet and the water box (Elliott Co.).

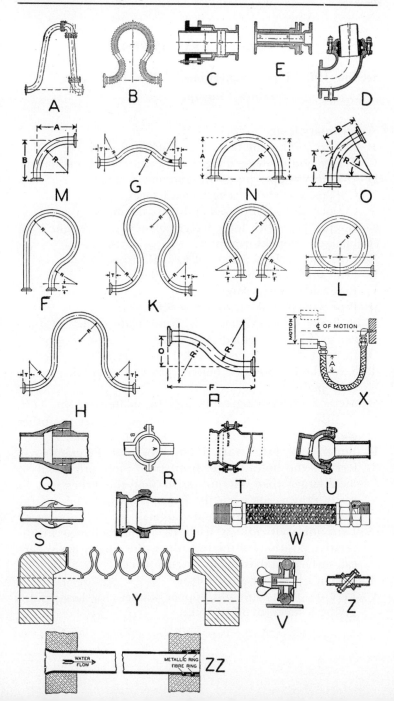

CLASS I.   FASTENERS

**Section 10f.   Pipe Connections, Boiler Tubes and
                Ferrules (Europe)**

A—Ordinary flange joint.

B—Flange joint with the pipes thickened near the flange.

C—Flange joint whose flange is strengthened by stiffeners.

D—Ordinary copper pipe with gun-metal flanges.

E—Copper pipe with strengthened flanges.

F—Copper pipe with brazed and riveted flanges.

G—Copper pipe with loose flanges of steel or wrought iron

H—Flanged wrought-iron pipe.

J—Pipe with welded flanges of wrought iron.

K—Mild steel, solid drawn, cast steel flanges.

L—Pipe with riveted flanges.

M—Pipe with shrunk-on rolled-steel flanges.

N—*A* Boiler tube expanded in front plate; *F* boiler tube
    expanded and ferruled on firebox.

O—Boiler tube expanded, beaded and ferruled on firebox.

P—Stay-tube for fire-tube boilers, preventing bulging of the
    tube plate.

Q—Method of screwing tubes into the plate; one end *A* is
    slightly enlarged to get a plus thread so that the other
    end *B* can be passed through the hole that fits *A*.

R—Boiler tube fitted with Humphrey and Tennant's cap
    ferrule; the holes in the front tube plate are about $\frac{1}{8}$
    inch larger than the tubes, so that the tubes can be
    easily drawn out from the front, the front ends of the
    tubes being correspondingly larger.

S—Lens joint; it adjusts itself to a small change in the
    relative positions of the two pipes; the ring *R* has spheri-
    cal surfaces and is made of gun metal.

T—Perkin's joint for small pipes.

U—Perkin's joint with soft copper washer, for large pipes,
    giving a metal-to-metal joint.

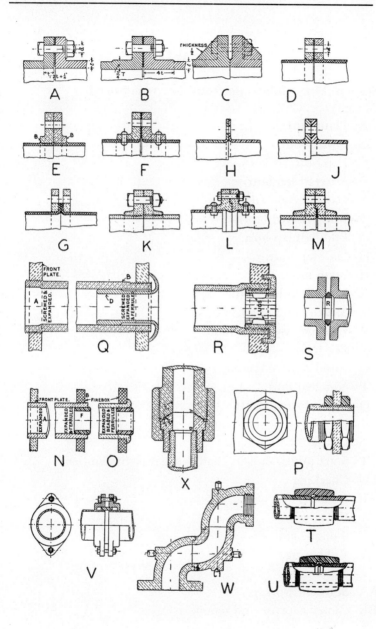

CLASS I.   FASTENERS

Section 11a.   Joints and Cuts for Woodworking

A—Plain butt.

B—Corner butt.

C—Lap butt.

D—Mitered lap butt.

E—Gained (housed) butt.

F—Blind-halved lap.

G—End half-lap butt.

H—Half-lap butt.

J—Plain miter butt.

K—Miter half-lap butt.

L—Mortise and tenon.

M—End mortise and tenon.

N—Peg tenon.

O—Blind housed tenon.

P—Notched butt.

Q—Plain dovetail butt.

R—Half-blind dovetail butt.

S—Dovetail half-lap joint.

T—Scarf joints, showing half joint at right-hand end.

U—Checking.

V—Chamfering.

W—Plowing.

X—Tongue and groove.

Y—Rabbeting.

Z—Stave work.

AA—Segment work.

BB—Closed mortise and tenon.

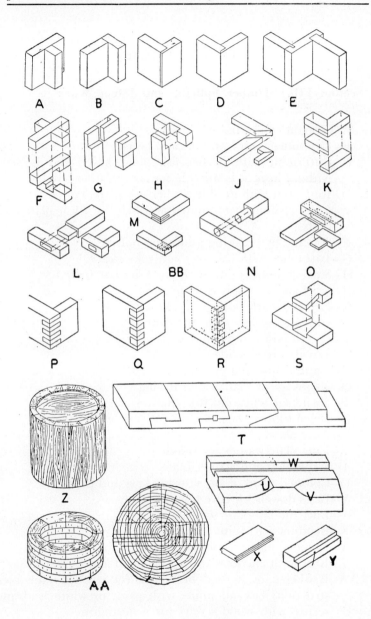

A   B   C   D   E

F   G   H   J   K

L   M   BB   N   O

P   Q   R   S

Z   T

W   U   V

AA   X   Y

CLASS I.   FASTENERS

Section 11b.   Timber Splicing and Joiner Work

A—Straight splice, bolted.

B—Lap splice with iron keys and bolts.

C—Butt joint with timber fish plate, keyed and bolted; without keys, it is suitable for compression only.

D—Butt joint with double timber fish plate, bolted.

E—Compression beam, bolted and held by a fish plate and bolts.

F—Splicing by breaking joints and bolting.

G—Lap splice with oak keys and yoke straps.

H—Scarf-and-butt joint, with one fish plate bolted.

 J—Scarf-and-butt splice, with iron fish plates bolted.

K—Lap-and-scarf butt joint, keyed with oak and locked with anchor fish plate and bolts.

L—Queen-post roof truss.

M—Wooden road bridge truss.

N—Natural splitting of tree.

O—Shrinkage of planks after drying.

P—Shrinkage when cut into quarters.

Q—Dried timber warped after planing on one side.

R—Fang plate.

 S—Fang-plate washer for wood.

T—Staple bolt and washer plate.

U—Colonial plank floor, showing keys and plugs.

V—Brass dowels for core boxes.

W—Butt-and-lap plate scarf joint.

X—Bending scarf joint.

 Y—Notched mortise-and-tenon joint with anchor strap.

 Z—Butt joint anchored with a key, bolt and washer.

AA—Rabbeted or housed buff joint for wooden tank construction; face all joints with paint or white lead and secure with wood screws for watertightness.

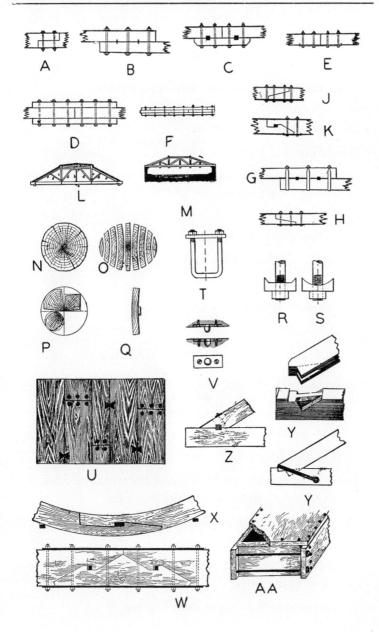

CLASS II.   ADJUSTING DEVICES

Section 12a.   Miscellaneous Adjusting Devices

A—Adjustable rod or lever.

B—Adjustable arm.

C—Adjustment for tension or compression of a torsion spring; the arm is split and locked to the spindle by a screw.

D—Adjustment of bearings for chain or belt gear.

E—Split joint for taking up wear.

F—Adjustable stays.

G—Adjustable turbine jet.

H, J—Adjustable table or base.

K—Slotted link and lock nut for adjusting the angle of a lever.

L—Disc and ring with partial angular adjustment by a screw and nut; used for self-centering chucks; the nut and bearing of the screw have allowance for swiveling.

M—Pin and hole adjustment for a lever.

N—Wedge bearing for locomotive horn-plate guides, slide bars and similar parts subject to wear.

O—Right- and left-hand screw and wedge adjustments for roller bearings.

P—Adjustment of engine crossheads for taking up wear on working faces.

Q—Adjustable floor-stand shaft bearing.

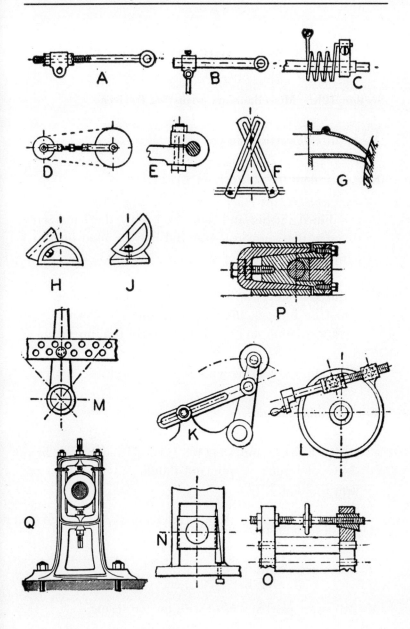

CLASS II.   ADJUSTING DEVICES

Section 12b.   Miscellaneous Adjusting Devices

A—Adjustable vertical sheave.

B—Adjustment for tailstock on a lathe.

C—Combined ratchet and hand-feed gear; the hand screw
turns in the worm-gear nut and may be used for quick
adjustment.

D—Adjustable step bearing with bronze bushing and step;
a mortise through the iron base and a key drawn with
screw extension and nut are for vertical adjustment.

E—Conical-pivot bearing with adjusting screw.

F—Adjustable hanger.

G—Collar bearing and step for vertical shafts; the thrust
sleeve of bronze is split and should have a key to prevent rotation.

H—Spiral torsion-spring adjustment.

J—Micrometer adjustment.

K—Adjusting pawl and head with torsion spring.

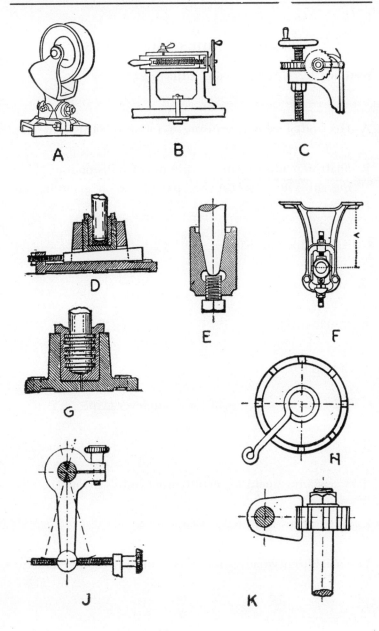

A

B

C

D

E

F

G

H

J

K

CLASS II.   ADJUSTING DEVICES

Section 12c.   Miscellaneous Adjusting Devices

A—Horizontal center adjustment for a vertical shaft step.

B—Shaft-step adjustment for spindles of millstones or grinding mills to regulate the space between grinding surfaces.

C—Adjustable post hanger.

D—Adjustable rack.

E—Ratchet rod.

F—Spring pawl.

G—Micrometer adjustment for a cam-lever grip.

H—Calipers.

J—Screw adjustment for maintaining rollers parallel.

K—Screw adjustment for a lever.

L—Variable-curve adjustment used in drawing instruments

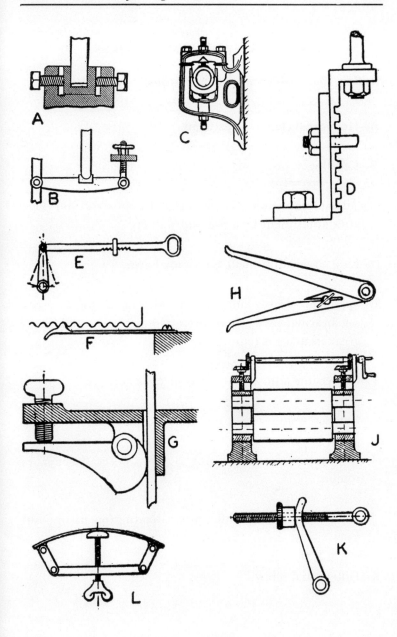

CLASS II.   ADJUSTING DEVICES

Section 13a.   Differential Screw-Adjusting Devices

A—Horizontal center adjustment.

B—Leveling adjustment.

C—Adjustable center pin traversed by a screw and fixed after adjustment by a nut and washer.

D—Fine screw adjustment for a radial arm.

E—Split cone-sleeves and setscrew adjustment for a revolving bearing, used where there is much wear and where great accuracy is required.

F—Center-line adjustment for lathe headstock, etc.

G—Fine screw adjustment for any movable part.

H—Adjustment for expanding a split borer, reamer or rosebit.

J—Division plate with differential dividing on its opposite faces.

K—Differential screw.

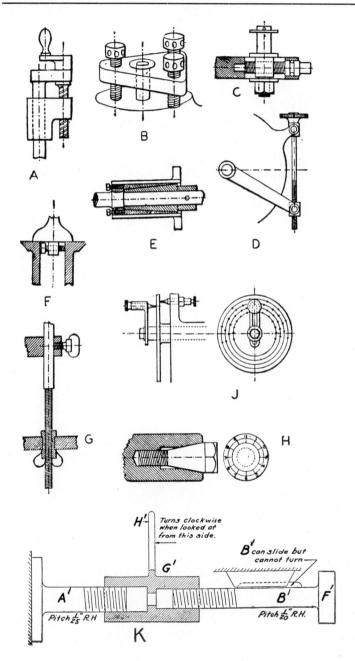

A

B

C

D

E

F

G

H

J

K

H'

Turns clockwise when looked at from this side.

G'

B' can slide but cannot turn

A'

Pitch $\frac{1}{25}$" R.H.

B'

F'

Pitch $\frac{1}{20}$" R.H.

CLASS II.   ADJUSTING DEVICES

Section 14a.   Valve Handles

A—Capstan handle for screw gear.

B—Ventilated twisted handle.

C—Dished hand wheel.

D—Bent looped-end handle.

E—Key handle for a cock.

F—Hand-wheel lock nut.

G—Hinged spanner for screwed glands.

H—Hand crank.

J—Locked hand wheels for valves to move in a certain order.

K—Faucet handle.

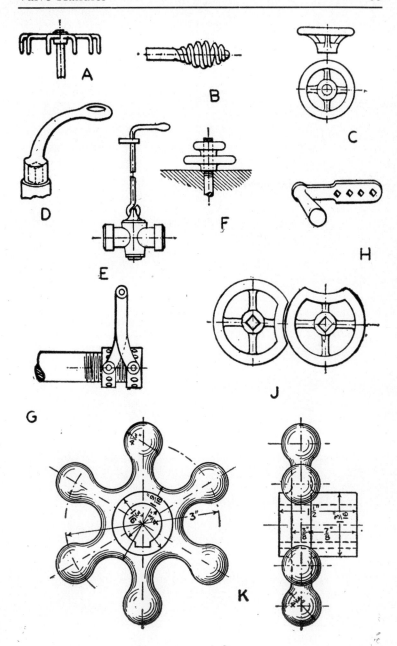

CLASS II.   ADJUSTING DEVICES

Section 14b.   Machine Handles

A—Typical boss and suitable for all handles B to J.

B—Forged lever handle.

C—Plain lever handle, flat type; it needs no machining.

D—Plain lever handle, clumsy, but comfortable to handle.

E to K—Lever handles most commonly used.

L—Stop-valve handle; it has no taper; used also for tap wrenches.

M—Box-spanner taper handle.

N—Stop-valve taper handle; used also for tap wrenches.

O—Box-spanner taper handle.

P—Stop valve or tap wrench in cramped position.

Q, R—Handles generally used on machine tools.

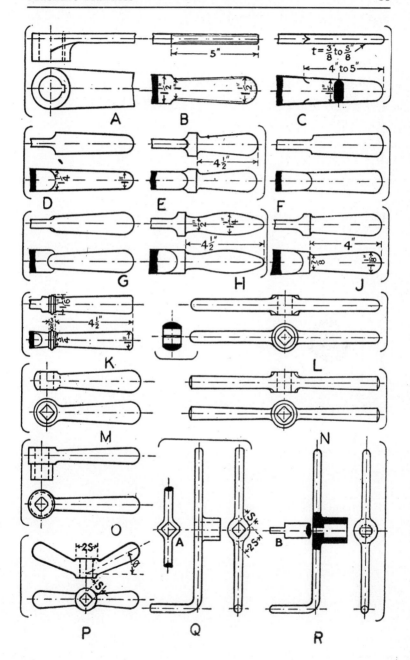

CLASS II.   ADJUSTING DEVICES

Section 14c.   Miscellaneous Handles

A—Hand wheel.

B—T-handle.

C—Capstan.

D—Bow or lifting handle.

E—T-bar.

F—Cross handle.

G—Knob.

H—Loop handle, hinged.

J—Loop handle, fixed.

K—T-handle.

L—Hand bar.

M—Swing-door handle.

N—Hinged lifting levers.

O—Bent handle for radial motion.

P—Stirrup handle.

Q—Loop handle.

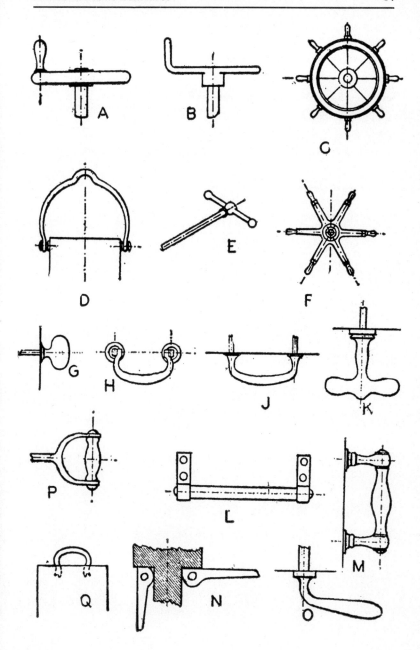

CLASS III.   SUPPORTS AND STRUCTURES

Section 15a.   Machine Frames and Bedplates

A—Shrunk ring fastening for cast-iron bedplates, wheels, etc.

B—Vertical columnar or distance-rod construction used for vertical compressors or engines and presses.

C—Foundation for a box bedplate.

D—Cast box bedplate.

E—Flat-bar or angle-iron side framing.

F—Base plate with concrete foundation.

G—Dovetail and key fixing for bracket bearings, etc.

H—Angle iron or flat bar.

J—Typical horizontal engine or compressor frame; heavy duty.

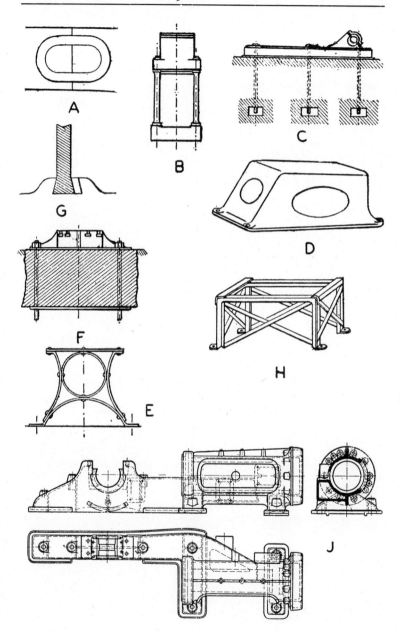

A

B

C

G

D

F

H

E

J

CLASS III.   SUPPORTS AND STRUCTURES

Section 15b.   Machine Frames and Bedplates

A—Rectangular section; simple but least economical.

B—Flanged or ribbed section; same weight as A, but much stronger.

C—Double I-section with web frame.

D—Hollow section.

E—Hollow section with tension flange.

F—Circular hollow section.

G, H—Circular hollow section with elliptical outline to support a bending action.

J—Rectangular bedplate; welded or riveted.

K—Open-box bedplate.

L—Closed-box bedplate.

M—Double-box bedplate with cross tie pieces.

N—Side-frame and distance-rod construction.

O—Side frames and cross bars on a base plate.

P—Table with four legs.

Q—Rectangular openwork box framing; used with cross shafts.

R—Sole plate and standard.

S—Wall bracket.

T—Steel side-plate and distance-rod construction.

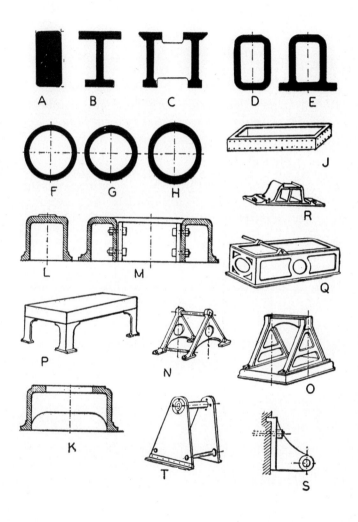

CLASS III.   SUPPORTS AND STRUCTURES

Section 15c.   Frames of Welded Design
(Lincoln Electric Co.)

A—Simple base plate.

B—Baling-press end frame with the main members composed of formed plate welded to form hollow sections.

C—End frame for a bending roll, having a built-up section of steel plate.

D—Support for a bearing which requires considerable clearance behind it; it is bolted to the base for maintenance purposes.

E—Bearing-block support and auxiliary arrangement.

F—Support for two bearings.

G—Base design of two different elevations.

H—Detail of the end of a base for holding down accessible bolts.

J—Slide.

K—Beam design.

L—Support for heavy loads.

M—Base detail.

N—Bearing support of simple construction.

O—Support for wheels.

P—Spacer design for a support.

Q—Spacer design for a gear case.

R—Parts used in Q.

S, T—Miscellaneous structures.

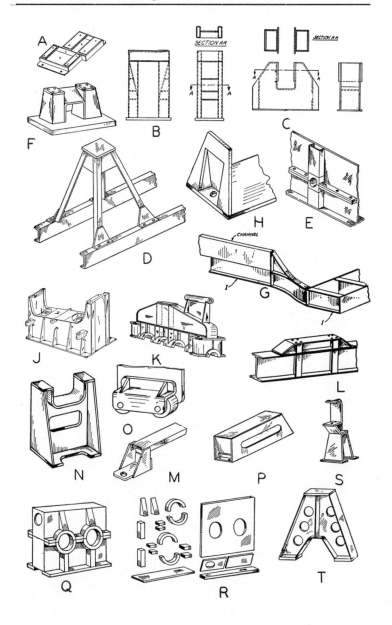

CLASS III.   SUPPORTS AND STRUCTURES

Section 16a.   Girders, Columns and Struts

A—Tubular swelled strut of steel plate; used for masts, sheer legs and crane jibs.

B—Strut formed of tube, with the end collars screwed in.

C—Ordinary solid swelled distance rod with collars; used for compressive strains.

D—Braced strut; usually of flat bars.

E—Double flat-bar cambered strut; stiffened by distance pieces and bolts.

F—Trussed strut; the trussing is 90° apart.

G—Steep roof truss; riveted design.

H—Box girder; riveted type; weight 10,617 pounds; supports at center 146,000 pounds.

J—Welded girder of the same design as H; weight 8,288 pounds; supports at center 146,000 pounds.

K—Craneway of welded construction; requires great lateral stability; tops of columns are connected by deep horizontal members (Lincoln Electric Co.).

L—Standard rolled shapes available for welded design.

M—Fabricated welded steel shapes.

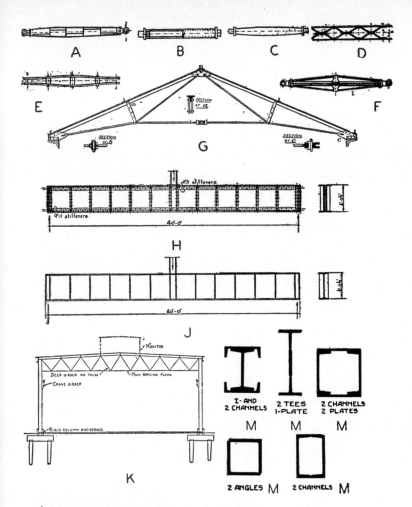

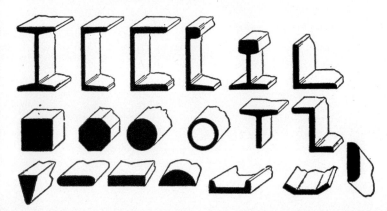

L - SHAPES AVAILABLE FOR WELDED DESIGN

CLASS III.   SUPPORTS AND STRUCTURES

Section 16b.   Bridge Trusses

A—Single-strut deck truss for short spans up to 40 feet.

B—Double-strut deck truss for spans up to 70 feet.

C—Multiple-strut deck truss for spans up to 100 feet.

D—Truss with interpanel tie rods (Whipple).

E—Truss in which the vertical and end posts are struts; it has vertical tie rods from the end posts and diagonal tie rods in the panels (Whipple).

F—Truss with vertical struts, except in the end panels which have vertical tie rods, with inclined end struts and diagonal tie rods.

G—Arch-deck truss bridge.

H—Truss with inclined strut and tie rod for each panel, with stiff compression upper chord, the vertical members being tie rods.

J—Truss with inclined posts and vertical tie rods (Baltimore model).

K—Arch-truss bridge.

L—Truss, having vertical end posts with inclined struts meeting at the center (Post).

M—Swing bridge (Whipple).

N—Swing bridge (Post).

O—Cantilever bridge.

P, Q, R—Suspension bridges.

S—Steel-arched concrete bridge (Thatcher type).

T—Rolling lift bridge (Chicago type).

U—Truss (combination of "Whipple" and "Warren System").

V—Brooklyn-New York bridge (Roebling suspension system).

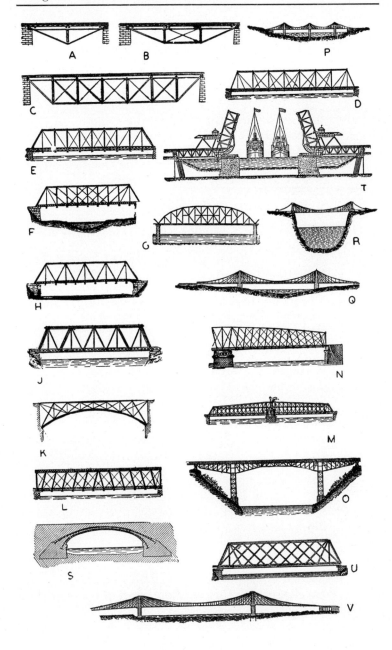

CLASS III.   SUPPORTS AND STRUCTURES

Section 17a.   Miscellaneous Bearings

A—Slot bearing for rising and falling spindle.

B—Hydraulic-oil pivot for vertical spindle, the oil be
ing forced under pressure into the channels and returned from the well to the pump.

C—Sliding bearing with vertical or horizontal travel.

D, E—Spherical and oval journals allowing shafts to run
out of line.

F—Plain double bearing with one cap and bolt.

G, H, J—Thrust bearings with collars and various retainers.

K, L—Ball and socket bearings for shafts considerably out
of line.

M—Balanced bearing to bear the weight of a light shaft
and to be placed between fixed bearings.

N—Half bearing; sometimes used without a cotter.

O—Double-V bearing for accommodating different
sizes of shafts.

P—Vertical pivot.

Q—Horizontal pivot and setscrew; the screw should
have a lock nut.

R—Conical neck with steel bush.

S, T—Swinging supports for a shaft having a sliding
bevel-gear or other motion; used for sliding gear
and overhead travellers.

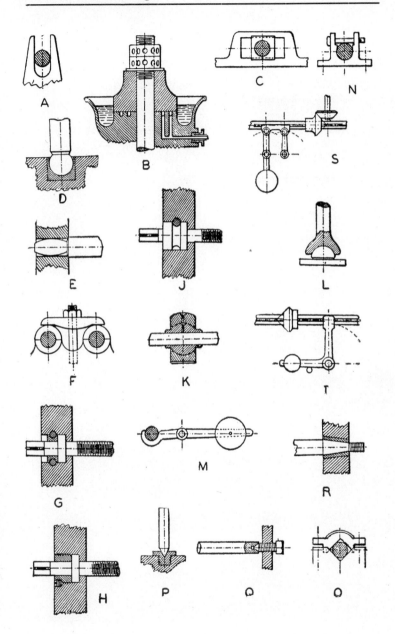

CLASS III.   SUPPORTS AND STRUCTURES

Section 17b.   Plain Bearings, Adjusting Brasses and Linings

A—Solid bearing; the simplest type with drilled hole.

B—Solid bearing with bushing.

C—Journal with solid collars.

D—Bearing cap for C.

E—Collar bearing for taking up thrust.

F—Thrust block.

G—Pivot, footstep or toe bearing for vertical shafts.

H to S—Various arrangements for varying pressure; in R, the wedge piece of the bolts for side adjustment has its thicker end at top which is objectionable; should the nut become loose, the wedge will work down and jamb the shaft.

T, U—Bushes lined with white metal.

V, W—White metal cast in holes.

X—White metal filling spiral grooves.

Y, Z—White metal filling rectangular recesses.

AA—Step entirely of white metal.

BB, CC—White metal fitted in solid strips driven-in like a key.

DD—Journal on shaft turned down.

EE—Journal on shaft with forged collar.

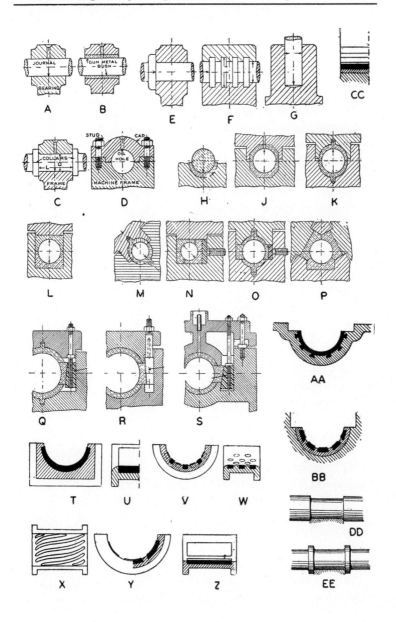

CLASS III.   SUPPORTS AND STRUCTURES

Section 17c.   Bearings, Journals and Hangers

A—Pivot bearings with slightly cupped steel discs whose rotation is blocked by a pin; the bronze bushing is prevented from rotating by snugs.

B—Pivot bearing with a hemispherical bronze disc block; rotation of the bush is prevented by the key *F* and that of the disc by the stop pin *P;* three oil grooves meet in the center of the disc.

C—Pivot bearing with four loose discs; oil is introduced at the center, and circulates by centrifugal force as shown by arrows.

D—Collar bearing; used at pressures above 200 pounds per square inch and up to 400 pounds per square inch with good lubrication.

E—Pivot with automatic lubrication by centrifugal force.

F—Schiele's pivot bearing which wears equally on all diameters; it is expensive to manufacture; it wastes energy in overcoming friction, but retains its shape as it wears and is self-adjusting.

G—Bearing with three brasses and top wedge bolts.

H—Vertical pivot with hardened screw.

J—Bearing for a rocker arm with tapered spindle; the wear is taken up by the screw.

K—Ball-bearing castor.

L—Bearing with two side brasses adjusted by screw.

M—Bearing with three brasses adjusted by screw.

N—Bearing with three brasses set up by side wedges and top screws; the bolts may become loose as the wedge tightens up by its own weight.

O—Adjustable intermediate bearing for a vertical shaft.

P—Pedestal bearing with four brasses.

Q—Center bearing with allowance for oscillation.

R—Hydraulic bearing whose shaft is sustained by water or oil pressure.

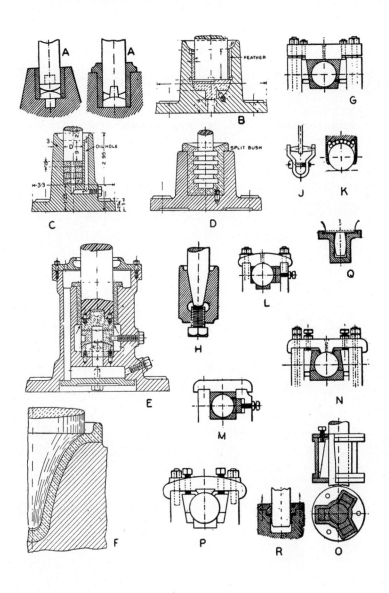

CLASS III.   SUPPORTS AND STRUCTURES

Section 17d.   Bearings, Journals and Hangers (German)

A—Transmission-shaft bearing; hanger type with ring oiling.

B—Typical pedestal.

C—Ceiling hangers.

D—Shaft bearing; hanger type; self-aligning.

E, F—Ceiling hangers.

G—Wall hanger.

H—Ring-oiling solid bearing.

J—Wall hanger.

K—Ring-oiling bearing and ring.

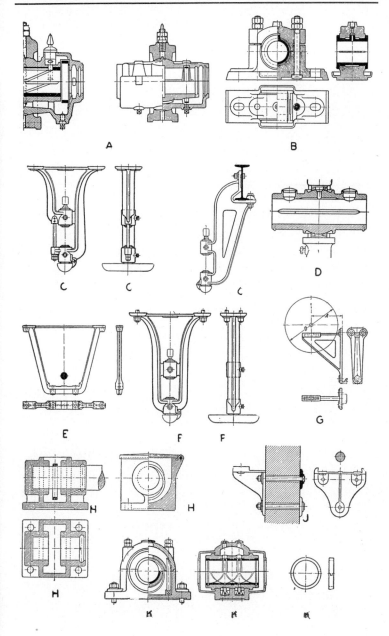

A

B

C

C

C

D

E

F

F

G

H

H

H

J

K

K

K

CLASS III.   SUPPORTS AND STRUCTURES

Section 17e.   Ball-Bearing Contacts

A—Simple two-point contact.

B—Grooved races, with their radius equal to that of the balls; excessive friction.

C, D—Examples of three-point contact.

E—Four-point contact; it will resist a certain amount of end thrust, but two firmly fixed bearings of this type should not be used on the same shaft; such a bearing, however, may be used with expansion-type bearings on the same shaft.

F—Race with radius $1\frac{1}{8}$ to $1\frac{3}{8}$ times that of the ball (first suggested by Stribeck).

G, H—Removable ring, permitting introduction or withdrawal of balls.

J, K—Journal-hub ball bearing; three-point contact; J shows cone.

L—Journal-hub ball bearing; four-point contact.

M—Cup and ball two-point contact for journal-hub bearings.

N, O—Three-point-contact thrust bearings.

P—Four-point-contact thrust bearing.

Q—Ball-thrust-bearing washer; two-point contact with cage (Hoffman type).

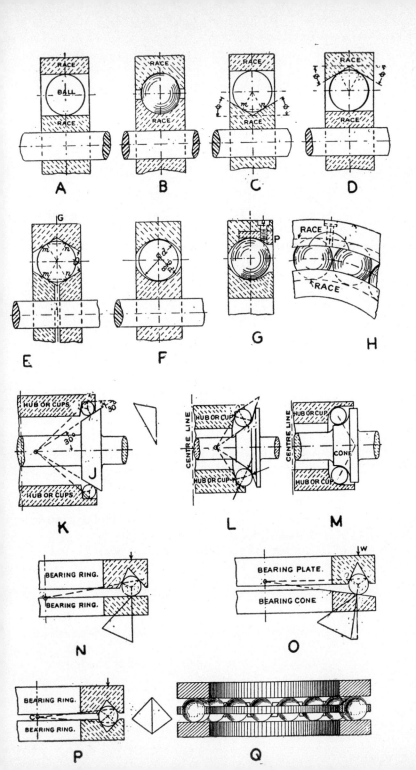

CLASS III.   SUPPORTS AND STRUCTURES

Section 17f.   Commercial Ball Bearings

A—Ball bearing with sleeve and spacing collars for thrust loads in alternate directions (Gwilliam).

B—Ball thrust bearing with leveling plates.

C—Grooved races without ball cage, using full circle of balls.

D—Thrust bearing designed to take slight radial load.

E—Thrust bearing with leveling plates, sleeve-spacing collars and single row of balls; used for thrust loads in alternate directions.

F—Ball thrust bearing with flat races and key slot.

G—Double-direction thrust bearing with leveling plates and spacing collars; the middle plate is locked to the shaft (Gwilliam).

H—Application of Garlock Klozure Oil Seals.

J—Conrad-type ball bearing for light radial loads; the inner race revolves, showing single shield (BCA).

K—Conrad-type ball bearing with felt seal (BCA).

L—Wide inner race with single labyrinth seal, set-screw burr-relief type; the inner race revolves (BCA).

M—Conrad-type ball bearing with double composition seal, single row (BCA).

N—Single-row angular-contact ball bearing; the load rating for pure thrust is 300% of radial load (BCA).

O—Ball bearing lubrication (New Departure Type).

P, Q, R, S—Lock nuts, lock washer and assembly (BCA).

T—Ball retainers for cup and cone bearings (BCA).

U—Ball retainers for thrust bearings (BCA).

V—Ball retainers for open-type radial bearings (BCA).

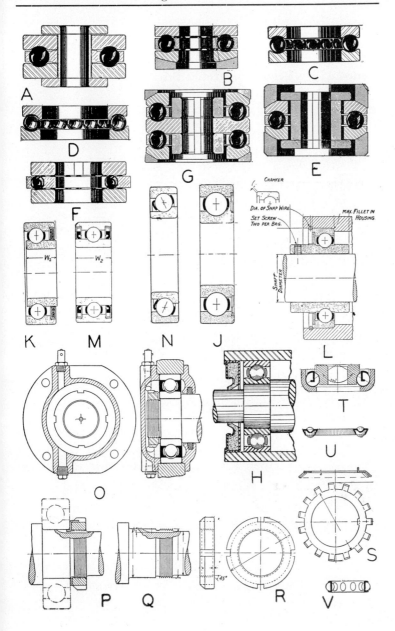

CLASS III.   SUPPORTS AND STRUCTURES

Section 17g.   Early-Type Roller Bearings

A—Simplest form of roller bearing.

B—Ring-cage roller bearing; an improvement over A.

C—Solid-cage roller bearing.

D—Hyatt flexible roller bearings (steel-bar winding); two shown.

E—Kynock flexible roller bearing (steel-plate winding); one shown.

F, G—Conical-roller thrust bearings.

H—Conical-roller thrust bearing with ball for heavy work.

J—Conical-roller thrust bearing complete with cage.

K—Cylindrical-roller thrust bearing and cage for heavy work.

L—Conical-roller thrust bearing.

M—Single-row straight bearing suitable for radial loads only.

N—Journal roller assembly for radial loads only.

O—Single-row tapered bearing for both radial and thrust loads.

P—Multiple-row tapered bearing for both radial and thrust loads.

Q—Needle roller bearing; no retainer ring; originally used for low-speed and oscillating parts.

R—Self-aligning bearing with hour glass rollers (Shafer type).

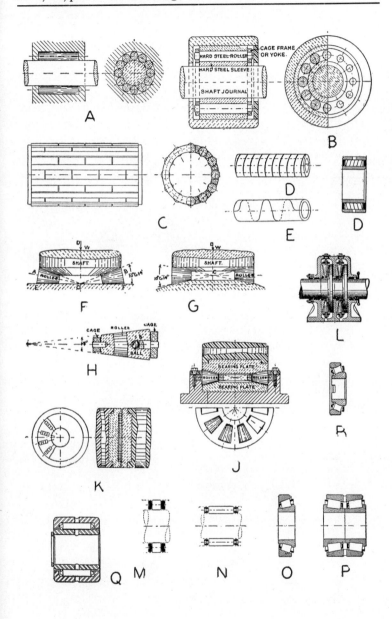

CLASS III.   SUPPORTS AND STRUCTURES

Section 17h.   Commercial Roller Bearings

A—Self-aligning flanged unit; nonexpansion type (Link-Belt).

B—Self-aligning flanged unit; expansion type (Link-Belt).

C—Direct shaft-end mounting with clamp plate (Link-Belt).

D—Self-aligning pillow block; closed end, expansion type (Link-Belt).

E—Self-aligning pillow block; open end, expansion type (Link-Belt).

F—Unmounted radial-thrust single-row roller bearing; angular-contact type; shims or threaded adjusting rings must be provided.

G, H—Unmounted double-row radial-thrust roller bearing (Link-Belt).

J—Plain roller thrust bearing combined with leveling plates and sleeve (Gwilliam).

K—Roller thrust bearing with concave and convex leveling plates; the concave plate is projecting on the sides.

L—Straight-sleeve general-purpose Timken roller bearing (Medart).

M—Tapered-sleeve Timken bearing mounted in a self-aligning pillow block; nonexpansion type; used for higher speeds and shock loads (Medart); one nonexpansion type bearing should be used as an anchor bearing and the balance of bearings should be of the expansion type to provide for lineal expansion and contraction of the shaft (Medart).

N—Self-aligning nonexpansion bearing for high-precision machines; the journaled shafts are shown; the inner races of bearings are pressed directly on the shaft (Medart).

O—Same as N, but showing the open end (Medart).

P—Rigid-nonexpansion type bearing for low speed and light duty.

Q—Same as N and O, but the shaft is not journaled.

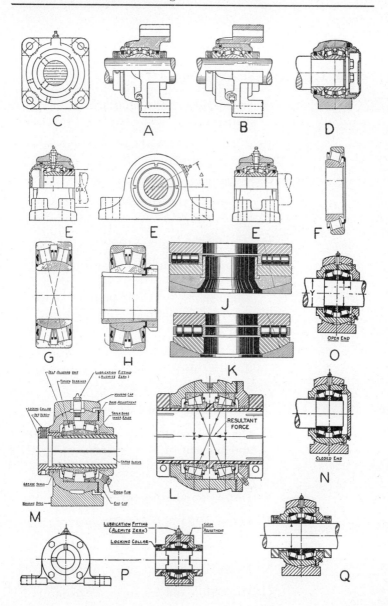

CLASS III.   SUPPORTS AND STRUCTURES

Section 17j.   Miscellaneous Supports

A—Stud-weld method of attaching Fiberglas insulation to ship sides or bulkheads.

B—Ratchet and weld-nail method of supporting insulation.

C—Typical hair-pin clip made of strap iron for supporting insulation to steel plating.

D—Steel-frame boiler support.

E—Grinnel preengineered spring hanger; available in fourteen sizes with a load range of 84 to 4,700 pounds; a dial is provided to indicate the cold and hot position load.

F—Sectional elevation of a cast-iron tank; showing fittings, and how supported.

G—Seven types of cylindrical tanks with dished and conical bottoms.

H—Detail of connection between the side and bottom of G.

J—Distortion and fractures; tension.

K—Distortion and fractures; compression.

L—Distortion and fractures; shear.

M—Distortion and fractures; bending.

N—Distortion and fractures; torsion.

O—Distortion and fractures; bearing.

P, Q, R, S, T, U, V—Stays, brackets for tanks, etc.

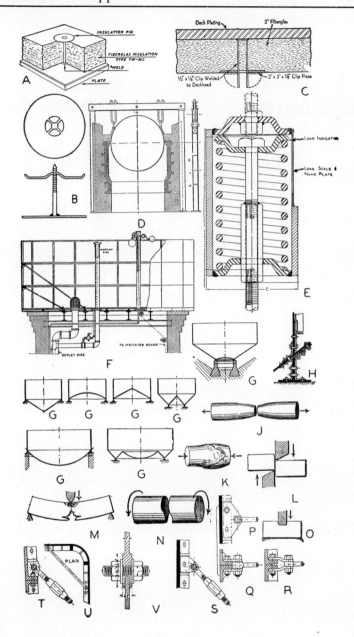

CLASS III.   SUPPORTS AND STRUCTURES

Section 17k.   I-Beam Supports and Tracks

A—Double I-beam overhead track.

B—Single I-beam overhead track.

C—Coburn-type track.

D—Bar track.

E--Monorail hoist.

F—Welded-type crane-rail supports.

G—I-beam track supports.

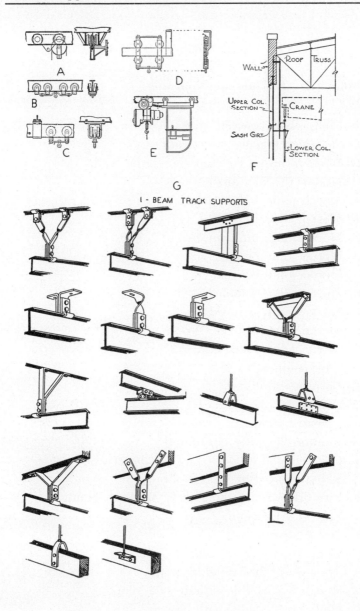

A

B

C

D

E

F

WALL
ROOF TRUSS
UPPER COL. SECTION
CRANE
SASH GIRT
LOWER COL. SECTION

G

I - BEAM TRACK SUPPORTS

CLASS III.   SUPPORTS AND STRUCTURES

Section 171.   Masonry and Concrete

A—Compressed concrete pile (Western Foundation Co.).

B—Sheathing bracing of excavations.

C—Wood forms for concrete foundation wall.

D—Hollow tile floor arches.

E—Mortar joint; weathered.

F—Mortar joint; struck.

G—Mortar joint; flush cut.

H—Mortar joint; concave.

J—Mortar joint; convex.

K—Mortar joint; V-tooled.

L—Mortar joint; stripped.

M—Mortar joint; rodded.

N—Mortar joint; raked out.

O—Ionic column.

P—Corinthian column.

Q—Doric column.

R—Roman column.

S—Reinforced concrete columns.

T—Concrete slab supported by steel beams.

U—Slab supported by concrete beams.

V—Face brick laid in common or American bond.

W—English brick bond.

X—Flemish brick bond.

Y—Stone work; coursed ashlar.

Z—Stone work; broken ashlar.

AA—Stone work; rubble with dressed joints.

BB—Stone work; rubble with broken joints.

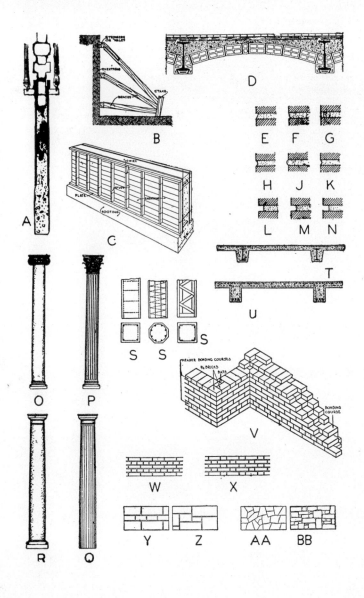

PART II—DYNAMICS

CLASS IV.   BASIC MECHANICAL MOVEMENTS

Section 18a.   Kinematic Chains

A—Link.

B—Slider crank chain.

C—Chain varieties:

1. Fixing AC gives direct-acting engine.

2. Fixing BC gives oscillating engine and quick return.

3. Fixing AB gives Whitworth's quick return.

4. Fixing block C gives Stannah's pendulum pump.

5. Fixing AC and prolonging CB to twice its length $(CB = AB)$ gives Scott-Russell's straight-line motion, D making a straight line.

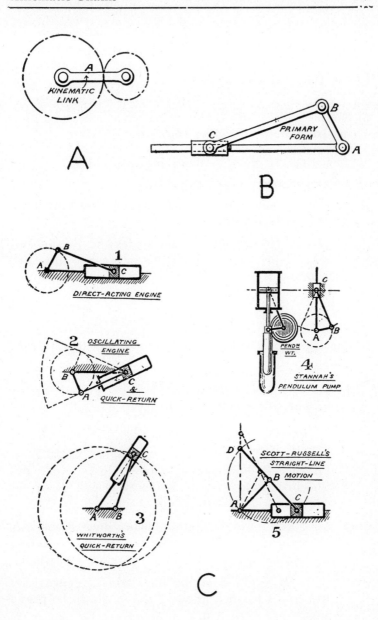

A

KINEMATIC LINK

B

PRIMARY FORM

1 DIRECT-ACTING ENGINE

2 OSCILLATING ENGINE

& QUICK-RETURN

3 WHITWORTHS QUICK-RETURN

4 STANNAH'S PENDULUM PUMP

PENDM. WT.

5 SCOTT-RUSSELL'S STRAIGHT-LINE MOTION

C

CLASS IV.   BASIC MECHANICAL MOVEMENTS

Section 18b.   Kinematic Chains

A—Double slider-crank chain; it has three links, two turning pairs and two sliding pairs variously connected:

   1. Fixing AC gives donkey-pump mechanism.

   2. Fixing AB and AC at right angles and removing the
       turning pair to C give elliptic trammels, oval chuck,
       and Oldham's coupling.

   3. Fixing AB and AC at right angles and putting one
       turning pair at C, two sliding and one turning pair
       at B give Rapson's slide which results in an increased
       leverage as the tiller is moved over hard.

B—Quadric-crank chain; it has four links and four turning
pairs:

   1. Fixing AB gives a beam engine with force closure by
       fly wheel lever-crank chain.

   2. Fixing AB and making AC equal BD give Watt's
       parallel motion.

   3. Fixing AB, but making opposite links equal give
       wheel coupling gears for locomotives with closure by
       double chain (parallel-crank chain).

   4. Same as 3, but altering the lengths gives special mo-
       tion; used in wire-rope manufacture for preserving
       the vertical position of drums (parallel-crank chain).

   5. Same as 3, but doubling the chain as shown gives
       Roberval's balance, allowing the weight to be placed
       anywhere on the pan.

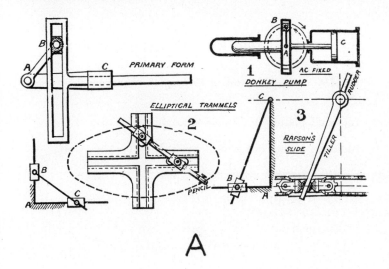

A

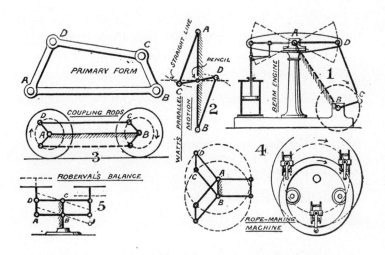

B

CLASS IV.   BASIC MECHANICAL MOVEMENTS

Section 19a.   Application of Energy to Machines

A—Inclined plane; W $\times$ sine of angle $+$ friction $=$ P.

B—Inclined plane; horizontal push, P $=$ W $\times$ h $\div$ b.

C—Wedge; strain $=$ force $\times$ l $\div$ W.

D—Screw; P $=$ W $\times$ pitch $\div$ 2 $\times$ r $\times$ 3.1416.

E—Worm gear; P $=$ W $\times$ pitch $\times$ r $\div$ 6.28 $\times$ r $\times$ R (W $\div$ 2, if the screw thread is double).

F—Chinese wheel, or differential axle; P $=$ W $\times$ (a—b) $\div$ 2 r.

G—Chinese windlass; the sheave and hook rise to a height equal to half the difference in the circumference of the barrels for each turn of the crank; forefather of the differential pulley.

H—Chinese shaft derrick; the bucket can be raised above the mouth of a pit or shaft.

J—Steam-hoisting engine, with reversing link.

K—Spanish windlass.

L—Two-ton-capacity jack screw.

M—Light-weight hoist over lathes and planers.

N—Capstan.

O—Winding engine with gear.

P—Steering gear for a boat.

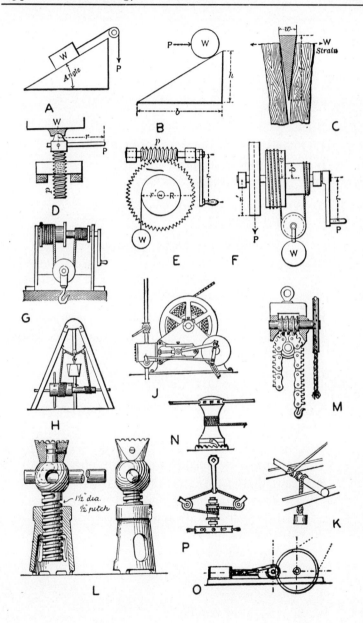

CLASS IV.   BASIC MECHANICAL MOVEMENTS

Section 20a.   Levers

A—Bell crank.

B—T or double-crank lever.

C—Plain lever with bosses for rod attachment.

D—Straight lever of the second order.

E—Foot-treadle frame.

F—Wrist plate or T-lever.

G—Double hand lever.

H—Lever formed with plates and distance pieces.

J—Hand starting lever; cheaply constructed with light channel iron; a bent lock bar, engaging holes in the sector plate, is cast on the bearing.

K—Double-lever hand motion for fire pumps.

L—Locking lever formed by an iron tube with a sliding catch rod inside the tube.

M—Convex worm for locking and adjusting a starting lever.

N—Equalizing levers for variable movements and springs.

O—Lever and rack lifting device.

P—Spring lever formed of steel plates.

Q—Compound lever; the roller board movements in the organs are similar, but each pair of arms and its shaft or roller is mounted independently on a pair of end centers.

R—Double lever for a plug cock.

S—Equalizing lever for two brakes.

T—Slotted-valve lever.

U, V, W—Starting-lever handles.

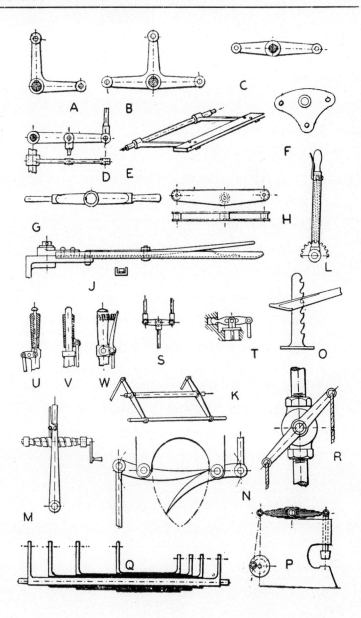

CLASS IV. BASIC MECHANICAL MOVEMENTS

Section 20b. Levers

A—First-order lever.

B—Second-order lever.

C—Third-order lever.

D, E, F—Plain levers with end bosses for rod attachments.

G—Forked-end offset lever.

H—Balance-weight lever.

J—Hand lever with round handle.

K—Starting lever with spring catch.

L—Headed lever for valve rods.

M, N—Rocking levers with sliding swivel joints.

O—Rolling washer and shot lever for a pull rod.

P—Ball-jointed levers; to allow for racking motion, the ball socket is split and bolted together.

Q—Hand lever of adjustable length.

R—Forked lever for spanning a central bearing.

S—Spring lever for locking in two positions.

T—Lever with universal motion.

U—Starting lever, with catch, hooked into holes in a sector plate.

V—Bell crank with plain end.

W—Foot lever.

X—Lever with spring catch.

Y—Plan of plain lever with forked ends.

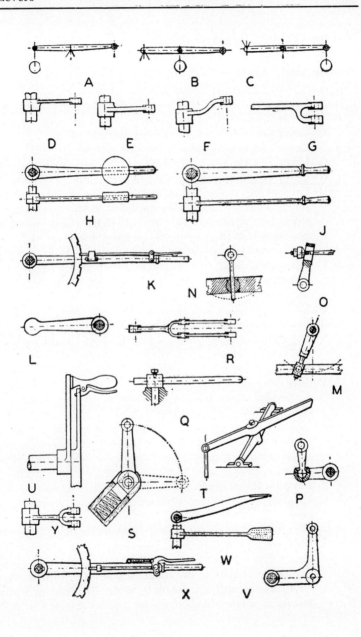

CLASS IV.   BASIC MECHANICAL MOVEMENTS

Section 20c.   Levers

A—Rope-twist bar.

B—Lever safety trip for a throttle valve.

C—Lever safety trip for a balanced throttle valve.

D, E, F—Lever and ratchet devices for cash registers.

G—Lever and ratchet for ringing the bell on a cash register.

H—Lever action applied to a typewriter roller.

I—Unhooking device; turning down the handle of the short bell-crank lever lifts the hook in the eccentric rod from the wrist pin of the rock-shaft crank to allow operation of the engine by a hand lever on the rock shaft; used on side-wheel steamers.

K—Compound-lever cutting pliers in which the toggle-joint principle is used to give greater power to the closing of the jaws.

L—Lever as applied to a Sheperd steam-engine governor.

M—Lever as applied to a Fitchburg steam-engine governor.

N—Toggle-joint lever press or punch; used in old printing and stamp presses.

O—Stump puller.

P—Tire shrinker.

Q—Suspended foot treadle; foot action prevents dead center of crank.

R—Steward mechanical compensating link davit; needs minimum effort to put out a lifeboat; *A* represents outreach, *B* shows inreach, *C* represents height from base to bite.

S—Landley sheath-screw lifeboat davit with straight boom.

T—Landley sheath-screw lifeboat davit with crescent-type boom.

U—Steward lifeboat release gear; releases both ends of the boat simultaneously with the throw of a single lever. Three types of hooks are available, e.g., type *"A"* hooks will release before the boat reaches the water; type *"AA"* hooks will release only when the boat is waterborne; or a combination my be used to release the stern first.

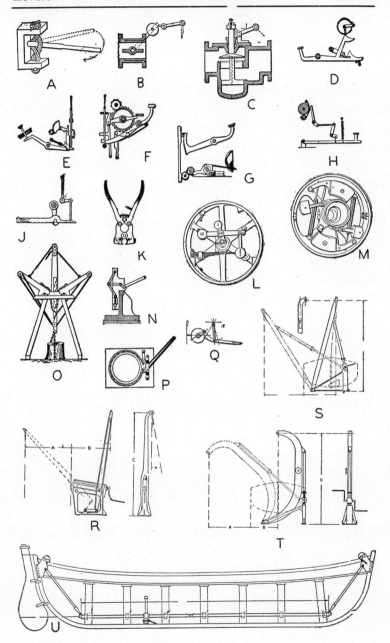

CLASS IV.   BASIC MECHANICAL MOVEMENTS

Section 21a.   Parallel and Straight-Line Motions

A—Peaucellier's parallel or straight-line mechanism; points A and B are fixed, point B being midway between A and C; point D is constrained to move in a straight line.

B—Watt's parallel motion for a beam engine.

C—Marine-engine, side-lever parallel motion.

D—Beam engine with rocking-link beam centers.

E—Vertical engine; radius bars A,A are pivoted to the frame at the middle of the stroke.

F—Direct-acting engine; the radius bars are of equal length from the center line of the engine and the sliding pivot of the long bar; at half stroke, both fixed and sliding pivots are at right angles to the center line.

G—Marine-engine side lever; a and b are of equal length; so are c and d; the length of the radius bar of rocker-shaft crank F equals b divided by e.

H—Cross-head slide for the vertical engine in a side-wheel steamer; obsolete design.

J—Parallel motion; all radius bars are of the same length as the half beam.

K—Parallel motion achieved by prolonging the piston rod and connecting it to the crank with a forked rod; old device used as pumps.

L—Forney's patent compensation weight for steam engines.

M—Parallel motion for vertical-acting engine.

N—Parallel motion; a equals b, c equals f, d equals e.

O—Parallel motion; a equals b, c equals d; e is a crosshead.

P—Parallel motion; c equals d, e equals c, b equals half of a.

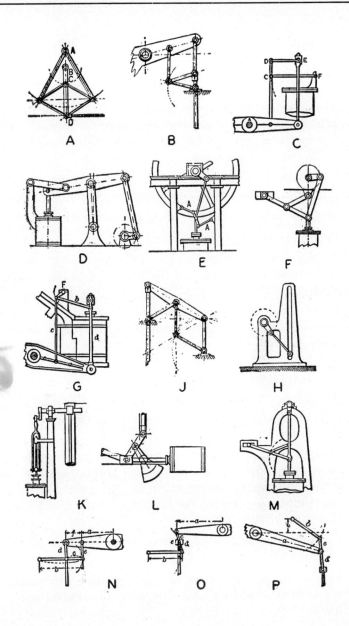

CLASS IV.   BASIC MECHANICAL MOVEMENTS

Section 21b.   Parallel and Straight-Line Motions

A—Beam-engine grasshopper motion, the slide being replaced by a long link; the gear may be formed with $AB = BC = BD$; or $AB : BC = BC : BD$; similar to D in the preceding section.

B—Grasshopper motion first used in a steam crane; the piston connects directly with D to the lift load.

C—Multiple grinding fixture, using parallel motion.

D—Parallel steering wheels.

E—Parallel motion of pliers; the jaws are double pivoted at equal distances from the central pivot of the handle.

F—Parallel motion; S and P are fixed; B moves vertically while L and M move horizontally parallel to fixed points S and P; links PO and OM are equal to BO; links LB and LM are of equal length, so are links SL and PD.

G—Double-link balanced scale.

H—Three-horse whiffletrees; the second pair has the center pins at two-thirds of their length from the inner end; the center single tree is attached with loose links.

J—Duplex air compressor with toggle joint; the crank moves the common joint of the long arm horizontally on a slide; the straightening of the toggle greatly increases the compression during the early part of the stroke when it is most required.

K—Pivoted steps for a gangway or accommodation ladder; the steps are always level.

L—Feathering paddle wheel or water motor; the paddles are kept vertical by a planetary gear; the central gear is fixed.

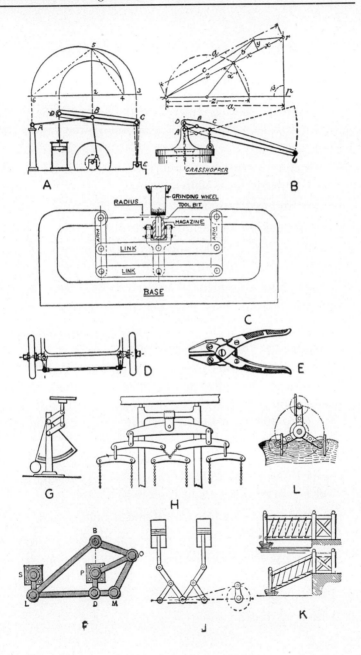

CLASS IV. BASIC MECHANICAL MOVEMENTS

Section 21c. Parallel and Straight-Line Motions

A—Beam with rocking fulcrum; AE and AC are equal when E is located on the center line of the piston rod; DB equals BC.

B—Two equal radius bars connected by a link having the main gudgeon in its center.

C—Sector and rack motion.

D—Parallel motion for an indicator pencil.

E—Parallel motion for an indicator, the approximate proportions being $c:d = d:b$.

F—Grasshopper movement of an early locomotive.

G—Parallel motion; guide-in-frame engine; very old.

H—Parallel motion achieved by a crosshead and rollers running against guide bars; very old.

J—Epicycloidal parallel motion; the pinion is one-half the diameter of the wheel at the pitch circle; the crank pin is fixed on the pitch circle of the pinion; the piston rod moves in a straight line; a curiosity of ancient engineering.

K—Cartwright's parallel motion; invented in 1787; both gears C are equal and also the two cranks A; the piston rod B moves in a straight line.

L—Beam with rocking fulcrum, A and A being equal.

M—Parallelly moving slides for hammers or other devices.

N—Floating table; it moves freely in any direction; it consists of two tables on two sets of rollers at right angles.

O—Parallel opening doors.

P—Universal drafting square.

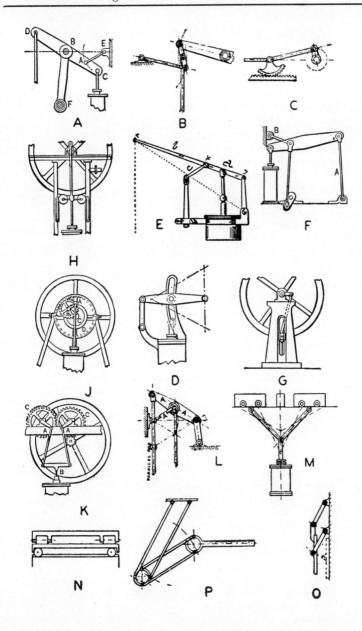

CLASS IV.   BASIC MECHANICAL MOVEMENTS

Section 22a.   Links and Connecting Rods

A—Turned and finished link without adjustments; the ends may be solid or forked.

B—Flat link of similar description as A, with raised bosses for facing and wear.

C—Adjustable link, with right- and left-hand screw coupling; lock nuts may be added to prevent the coupling from working loose.

D—Strap link fitted with brasses, gibs and cotters and distance bar; the wear of brasses is all taken up in one way by the gib and cotter; therefore, if great accuracy in the distance from centers is necessary, gibs and cotters should be fitted at both sides of one pair of brasses.

E—Turned link with adjustable end brasses; the forked end should be used where there is the greatest wear.

F—Wood connecting rod or pump rod with wrought-iron strap ends.

G—Typical shifting link for link-reversing gear.

H—Link similar to G, but having suspension of a side pin.

J—Reversed-curve link.

K—Solid-bar link; the valve rod and eccentric rods have forked ends.

L—Double-bar link.

M—Attachment for connecting a rod to a pump ram.

N—Connecting-rod end, the rear brass being held by a setscrew with coned point, which penetrates and displaces a number of steel balls.

O—Wedge cotter and brass bearing.

P—Antifriction-rod end for applications where strain is all on one stroke (as in single-acting pumps); the strain is acting on a friction roller.

Q—Simple connecting-rod end and half brass for single-acting pumps.

R—Solid link with swiveling segments in the box on the valve spindle.

S—Solid end rod split with screw-bolt-tightening device.

T, U—Solid ends for small rods.

V—Common forked rod end with cap.

W—Hook-bolt attachment for crosshead pin.

X—Double connecting rod in which the rods form also distance rods and bolts for the heads, which are in valves and fitted brasses of the ordinary type.

Y—Plain link.

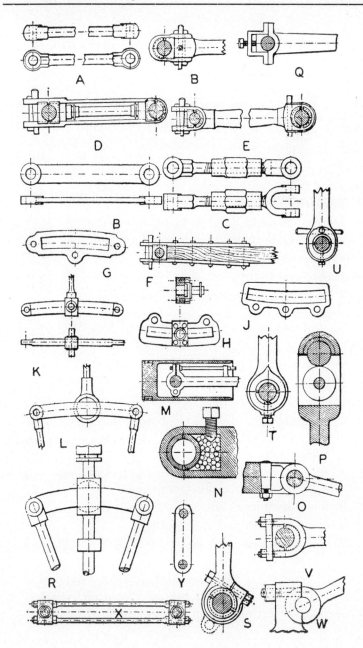

CLASS IV. BASIC MECHANICAL MOVEMENTS

Section 22b. Connecting Rods

A—Marine connecting-rod head, in which the brasses are extended to form the central block in halves, the rod end being T-shaped.

B—Connecting-rod end; the end block can be taken out sideways.

C—Solid end for connecting rod; the brasses can be adjusted by capstan screw.

D—Marine connecting-rod end with half brass and brass cap.

E—Solid-end rod; the brasses are slipped in sideways; locked with cotter and setscrew.

F—Forked-end rod.

G—Strap-head connecting rod with screw cotter.

H—Covered solid end for crank pins with screw adjustment for the brasses.

J—Strap end for heavy rods with cotter for tightening the strap to the V's in the connecting-rod end.

K—Rod end with side strap.

L—Solid end with double setscrew fastening for the cotter.

M—Connecting-rod end, crosshead and crosshead pin, showing metallic renewable plugs in the wearing faces of the crosshead pin.

N—Strap end with diagonal key; more accessible than a straight key under certain conditions.

O—Rod end with fixed pin secured by nut and cotter.

P—Rod end with hinged strap and bolt.

Q—Simple connecting-rod end half brass for single-acting pumps or compressors.

R—Yoke connection for a continuous piston rod and outside cylinder; crank shaft beyond the steam cylinder.

S—Spring connecting rod of steel or wood.

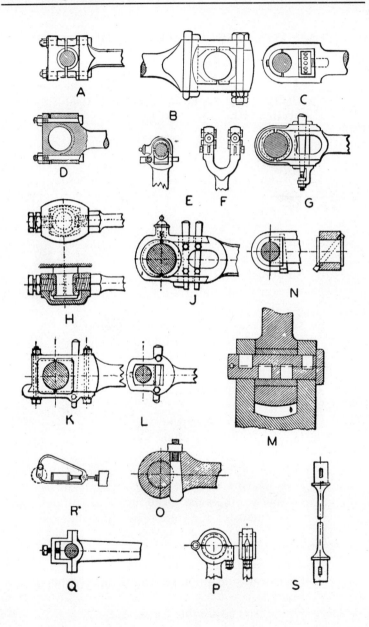

CLASS IV.   BASIC MECHANICAL MOVEMENTS

Section 22c.   Guides and Slides

A—Two bars and crosshead; the bars must be suffi-
ciently apart to allow for the angle of the connecting rod.

B—Bar and slipper.

C—Four bars, crosshead and slide blocks; the connecting rod works between two pairs of guides; the
bottom guides are often cast solid with the bedplate.

D—Adjustable slipper; there are other adjustments
for wear by wedges.

E—Section of C and alternative crosshead for two
round bar guides.

F—Slide bed and slipper.

G—Section of trunk guide cast with the engine bed
and bored out.

H—Oscillating-cylinder piston-head guides.

J—Oscillating fulcrum instead of guides.

K—Diagonal crosshead and guide bars to allow the
crank and connecting rod to pass the guide bars.

L—Round bar and flat guide bed.

M, N, O—Valve-rod guides.

P, Q—Guide rollers for ropes.

R, S—Guide rollers for bars of various sections.

T—Vertical bracket-cage guides.

U—Crosshead for two single-bar guides with renewable wearing strips and square guide surface.

V—Crosshead single-bar guide, with or without lower
attachment for the compressor or pump rod.

W—Double V-guide for a crosshead.

X—V-guide bar and guide.

Y—Crosshead bent around the rod.

Z—Crosshead side guide for an engine, compressor or
pump.

AA—Guide bars adjustable for wear.

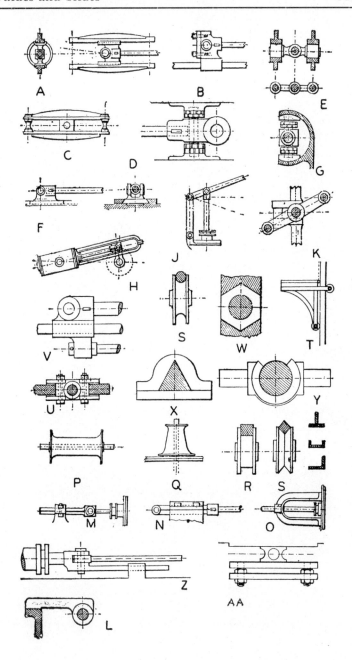

CLASS IV.   BASIC MECHANICAL MOVEMENTS

Section 22d.   Guides and Slides

A—Double V-bed with setscrew adjustments.

B—Guide bed for planing or other machines in which the bed is not raised by weight.

C—Deep V-guide; used for crossheads, tool boxes, etc., requiring accurate movement.

D—Lathe bed with square guides and adjustments for wear.

E—Double V-bed for planing machine.

F—Radial slide for a tool box.

G—Engine crosshead, formed of two slide blocks cast in one piece with the crosshead pin and two caps bolted together enclosing the piston-rod end and bolted to the slide blocks.

H—V-guides with bevelled adjusting strip and setscrews.

J—Crosshead guide, with two slide bars.

K—Crosshead guides of square section.

L—V-guides with a loose V-strip set up by screws on top.

M—Simple guide attachment to a plain bar.

N—Guide bed with square guides and renewable strip adjusted by setscrews.

O—V-guide with V-strip and setscrew adjustment.

P—V-guide adjusted at the top.

Q—V-guide with bevelled strip and setscrew.

R—Engine crosshead with adjustable guide brasses set up by a taper key and nuts.

S—V-guide with setscrew adjustment.

T—V-guides, with renewable strip.

U—Curved-segment guide for a link movement to turn an angle.

V, W—Rope guides.

X—Sliding bed guided by two square-grooved strips; adjusted by diagonal setscrews.

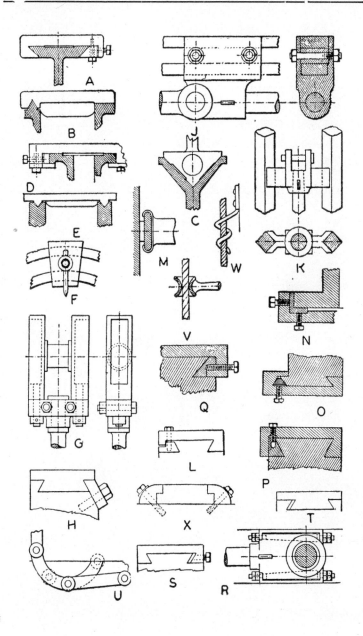

CLASS IV.   BASIC MECHANICAL MOVEMENTS

Section 23a.   Blocks and Tackles

A—Differential pulley; $W \div P = 2R \div R - r$ is the mechanical advantage.

B—Block and tackle; it has a mechanical advantage of 2 for each movable pulley; in this case $3 \times 2 = 6$; $P = W \div 6$.

C—Single whip; $P = W;$* $\quad \dfrac{P}{W} = \dfrac{11}{10}$ **

D—Single whip with block at weight; $\dfrac{P}{W} = \dfrac{10}{20}; \dfrac{P}{W} = \dfrac{12}{20}$

E—Gun tackle purchase; $\dfrac{P}{W} = \dfrac{10}{20}; \dfrac{P}{W} = \dfrac{12}{20}$

F—The same inverted; $\dfrac{P}{W} = \dfrac{10}{30}; \dfrac{P}{W} = \dfrac{13}{30}$

G—Luff tackle; $\dfrac{P}{W} = \dfrac{10}{30}; \dfrac{P}{W} = \dfrac{13}{30}$

H—The same inverted; $\dfrac{P}{W} = \dfrac{10}{40}; \dfrac{P}{W} = \dfrac{14}{40}$

J—Double purchase; $\dfrac{P}{W} = \dfrac{10}{40}; \dfrac{P}{W} = \dfrac{14}{40}$

K—The same inverted; $\dfrac{P}{W} = \dfrac{10}{50}; \dfrac{P}{W} = \dfrac{15}{50}$

L—Spanish Burton; $\dfrac{P}{W} = \dfrac{10}{30}; \dfrac{P}{W} = \dfrac{13}{30}$

M—Double Spanish Burton; $\dfrac{P}{W} = \dfrac{10}{50}; \dfrac{P}{W} = \dfrac{15}{50}$

N—Bell purchase; $\dfrac{P}{W} = \dfrac{10}{70}; \dfrac{P}{W} = \dfrac{17}{70}$

O—Luff upon luff; $\dfrac{P}{W} = \dfrac{10}{160}; \dfrac{P}{W} = \dfrac{26}{160}$

*In C to O, the first equation shows the ratio of power to weight friction not considered.

**In C to O, the second equation shows the ratio of power to weight friction considered.

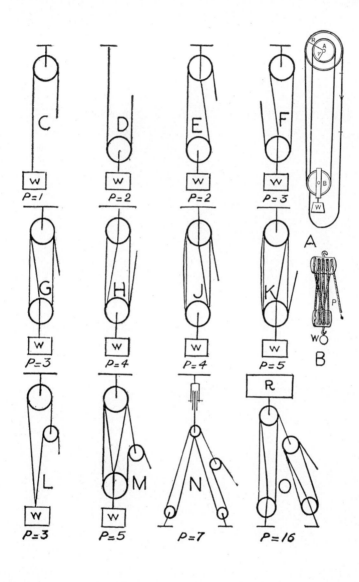

CLASS IV.   BASIC MECHANICAL MOVEMENTS

Section 23b.   Differential Pulleys and Winches

A—Moore's differential pulley block.

B—Weston's differential pulley block; consists of a two-grooved pitched chain sheave having different numbers of teeth, in combination with a return block and end-less chain.

C—Self-sustaining pulley block.

D—Differential pulley block (German).

E—Differential pulley block for light work (German).

F—Hoisting drums and quadruple block (German).

G—Electric hoist (German).

H—Hand hoist (German).

J—Travelling hoist (German).

K—Jack (German).

L—Yale and Towne screw-geared chain hoist section.

M—Chain offset (German).

N—Chain sheaves and fittings (German).

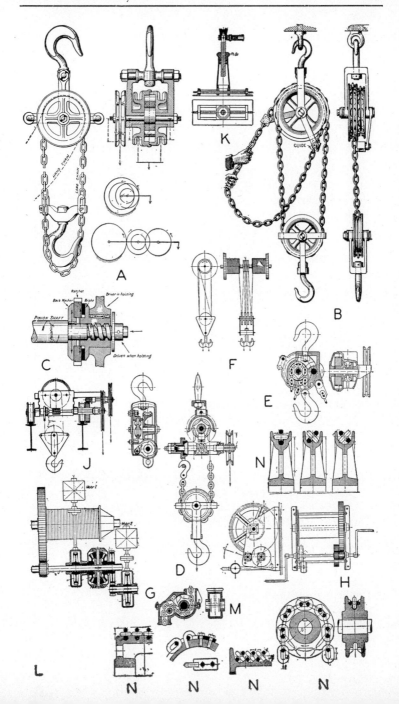

CLASS IV.   BASIC MECHANICAL MOVEMENTS

Section 23c.   Chains, Crane Hooks and Fittings

A—Links of ordinary close chain, short link, rigging or crane chain.

B—Oval stud link with broad-headed stud.

C—Oval stud link with pointed stud.

D—Parallel-sided stud link.

E—Obtuse-angled stud link.

F, G—Scarfed joints.

H, J, K—The effect of wear on links.

L—Spring catch which prevents the chain from working loosely.

M—Ball-bearing hook which reduces friction in swiveling.

N—Foundry-charge hook.

O—Ramshorn or double-crane hooks.

P—Ordinary swivel hook.

Q—Ramshorn or double hook.

R—Single-sling or lashing chain.

S—Shackle and swivel.

T—Stopper for riding bits (Blake).

U—Ordinary cable swivel.

V—Chain with shackle.

W—Screwed messenger link.

X—Screwed D-shackle (ordinary chain type).

Y—Cotter messenger link.

Z—Francis split link.

AA—Split link.

BB—Harp shackle with forelock (anchor type).

CC—Screwed connecting link.

DD—Chain tighteners or screws for sling chains.

EE, FF—Chains allowing alternate links to lie flat on the sheave.

GG, HH—Chain sheaves with curved groove.

JJ—Variety of GG.

KK—Cargo hook.

LL—Shunting hook (Liverpool type).

MM—Locking device.

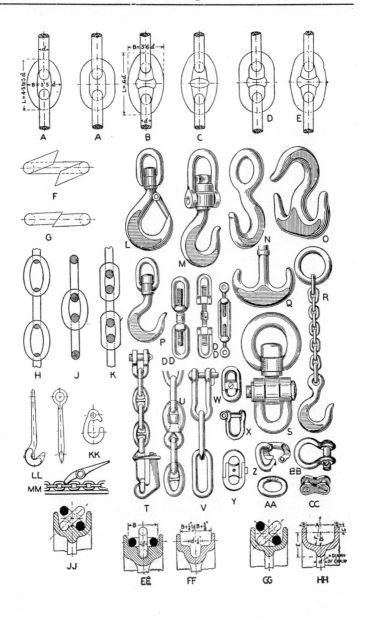

CLASS IV.   BASIC MECHANICAL MOVEMENTS

Section 23d.   Hooks and Swivels

A—Fixed-bar hook with snap.

B—Snap hook.

C—Slip hook with mousing ring.

D, E—Slip hooks.

F—Double or match hooks.

G—Self-locking hook with pin and inclined shoulder.

H—The common "Lewis."

J—Self-gripping claw grab.

K—Grab bucket.

L, M, N—Double S-links.

O—Snap link.

P—Automatic slip hook.

Q—Self-locking draw-bar hook.

R—Safety link; it has a flat on the link to slip through a notch.

S—S-link.

T—Split link.

U—Hook with rope grip.

V—Split link.

W—Hook eye for guy rope.

X—Slip hook for pile driver.

Y—Loop or eye shackle.

Z—Swivel-snap hook.

AA—Swivel shackle.

BB—Spring-snap hook.

CC—Grip hook.

DD—Slip hook for towing.

EE—Towing hook with mousing rope.

FF—Swivel-snap hook.

GG—Double link and bolt connection.

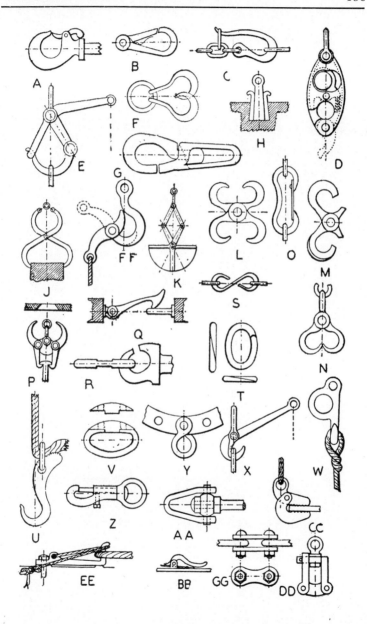

CLASS IV.   BASIC MECHANICAL MOVEMENTS

Section 24a.   Cams, Tappets and Wipers

A, B, C—Heart cams for giving regular or intermittent motion to a follower.

D—Crown cam for vertical shaft.

E, F—Jumping cams.

G—Covered heart cam.

H—Covered crown cam.

J—Wiper and lever action.

K—Twisted bar with sliding bush causing the bar to turn.

L—Cam plate and levers with rocking motion; can impart any intermittent or variable motion to a follower.

M—Cam-lever motion from a reciprocating rod to give irregular motion to another rod.

N—Crank pin and slotted lever; this combination gives variable speed with quick return.

O—Spiral or wheel cam.

P—Compound cam used for operating a number of radial grips or arms.

Q—Internal compound cam used for operating several radial slides or internal holders.

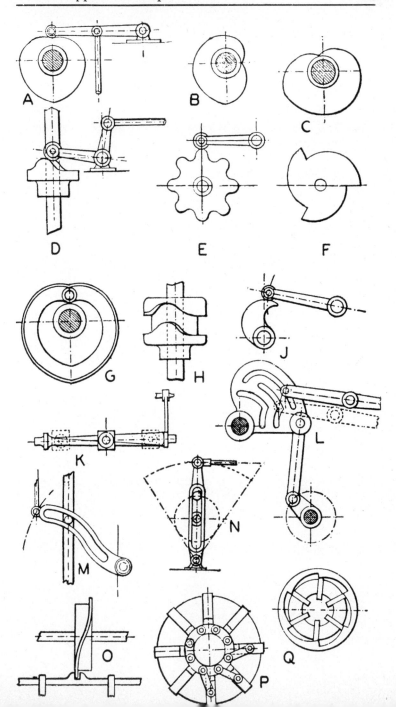

CLASS IV.   BASIC MECHANICAL MOVEMENTS

Section 24b.   Cams, Tappets and Wipers

A—Diagonal-disc cam or swash plate.

B—Belt shifter.

C, D—Sectors and bent lever used on steam-engine valve gear (old).

E—T-lever valve motion; used on compressed-air drills and steam engines.

F—Four-bolt camplate; used for screwing dies and locks.

G—Slot-cam and lever motion.

H—Barrel motion for musical instruments, looms, etc.

J—Drum with spiral vanes of long pitch; used for intermittent circular motion.

K—Volute and lever.

L—Double screw for converting circular into reciprocating motion; it has a right- and left-hand screw thread.

M—Eccentric-ring and roller motion for converting circular into reciprocating motion.

N—Stamp mill.

O—Scroll cam.

P—Crank and lever for intermittent or continuous motion.

Q—Piston or valve-rod lever motion.

R—Similar motion to O with a roller at the end of the lever.

S—Rod and lever reciprocating motion with a roller at the end of the lever.

T—Similar motion to S with a socket forged into the rod.

U—Triangular cam; it gives three reciprocations in one revolution.

V—Cam for giving motion to several rods simultaneously; used in organ pedals.

W—Crossed-lever motion.

X—Adjustable crown cam and roller.

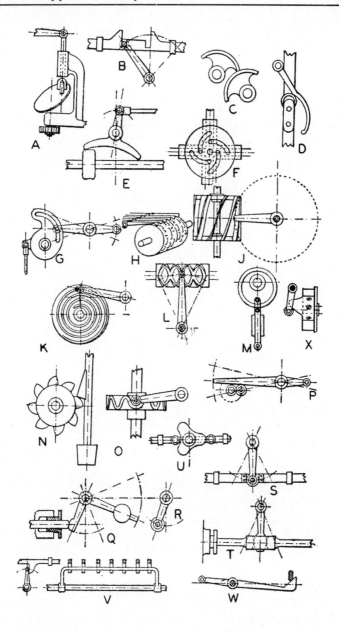

CLASS IV.   BASIC MECHANICAL MOVEMENTS

Section 24c.   Cams, Tappets and Wipers

A—Covered drum cam whose parts can be replaced to give a different movement.

B—Diagonal disc cam, giving a rocking movement to the crosshead and shaft.

C—Spiral radius bar for opening a valve which is lifted off its seat by the radial motion of the lever against the inclined radius bar.

D—Crank-pin and slotted lever motion with the slot arranged for irregular or intermittent motion.

E—Eccentric and slotted arm; the pin at the top of the arm has both a vertical and horizontal motion, causing it to trace an ellipse; the pin on which the slot runs is fixed.

F—Wiper and lever motion with rubbing plates; used on old engines.

G—Allen valve lift or toe operated from a cam on a rock shaft.

H—Cam groove and rocking arm.

J—Yoke strap and eccentric circular cam.

K—Triangular curved eccentric with a stop at each half revolution.

L—Triangular eccentric; similar to K.

M—Reciprocating motion with four stops, two of which are of longer duration than the others.

N—Uniform reciprocating motion derived from the circular motion of a disc or crank; the endless groove in the crosshead conforms to the varying rectilinear motion of the crank pin.

O—Needle-bar slot cam for sewing machines; the depression in the pin slot gives the needle a stop while the shuttle passes.

P—Variable crank throw operated by the slotted sector on the face plate.

Q—Variable rectilinear motion of a shaft derived from a vibrating curved, slotted arm.

R—Beveled disc cam for imparting variable reciprocating motion to a bar at an angle to a shaft.

S—Closure of rollers.

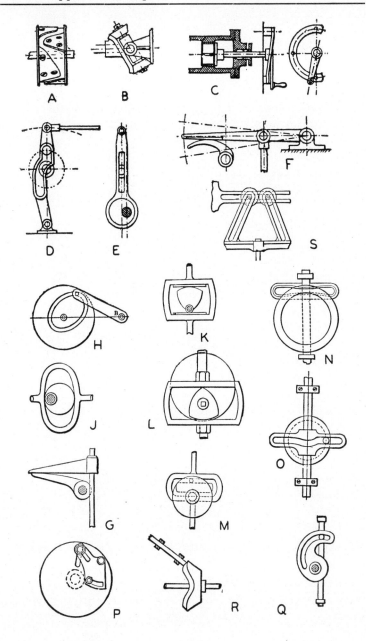

A

B

C

D

E

F

S

H

K

N

J

L

O

G

M

P

R

Q

CLASS IV.   BASIC MECHANICAL MOVEMENTS

Section 24d.   Cams, Tappets and Wipers

A—Variable crank pin; a slotted base plate backed by a spiral slotted plate.

B—Principle of cam application.

C—Double cam motion from a sliding follower.

D—Heart cam; grooves in the face plate vibrate a lever which gives an irregular swinging motion.

E—Irregular reciprocating motion.

F—Rotary motion of a three-arm wiper produces reciprocating motion.

G—Power escapement for heavy machines.

H—Irregular vibrating circular motion from the continuous circular motion of a cam slot.

J—Vibrating rectilinear motion from a revolving trefoil cam, operating on a tappet.

K—Irregular cam motion to operate valve rods used on old steam engines.

L—Cam-operated shears.

M—Scroll cam.

N—Clover-leaf cam for rollers to give them a bearing in all positions.

O—Reciprocating rectilinear motion.

P—Gear-disengaging cam lever.

Q—Cam sectors of logarithmic spiral wheels; the sum of the lengths of every pair of coincident radii is always equal to the distance of centers from each other.

R—Equalizing levers or toes for variable rod movement.

S—Angular wipers for operating valves of beam engines.

T—Wiper cam for stamp mill.

U—Bell-crank toe levers.

V—Cam-lever grip for rod or rope stop; this principle is used on safety catches for elevators.

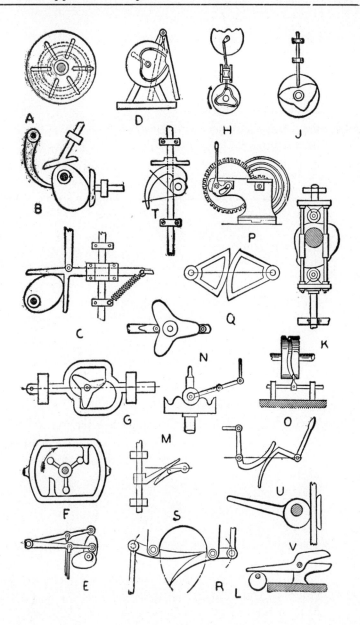

CLASS IV.   BASIC MECHANICAL MOVEMENTS

Section 24e.   Built-Up Gear Wheels

A, B—Built-up gear wheel; an expensive method of connect
ing segments of a rim.

C, D—Built-up gear wheel; a common method of fixing cast
iron segments of a rim.

E, F—Built-up gear wheel; method of joining an arm to the
rim by means of wedges, the spaces between the jog-
gles being filled up with dry oak and wedged up with
iron wedges after the wheel is in place; strengthening
hoops are shrunk on each side.

G—Built-up gear wheel; methods of attaching teeth castings.

H—Built-up gear wheel; method of fixing steel segments.

J—Built-up gear wheel; method of joining segments together at the rim and joining the boss at parts between the arms.

K—Stepped toothed gearing; invented by Dr. Hooke.

L, M—Built-up gear wheel; similar to J.

N, O—Built-up gear wheel; method of attaching arms to rim;
teeth may be attached as in G.

P, Q—Built-up gear wheel; method of attaching rim.

R, S—Built-up gear wheel; end of arm has a projecting
piece *A* let into the rim to take the shearing stress.

T, U—Built-up gear wheel; key made in two pieces driven
in from opposite sides.

V, W—Built-up gear wheel; for very large wheels, where boss
is cast separate from arms and segments; arms are
keyed in.

X—Built-up gear wheel; alternative method of fixing
arms to boss by bolting on; used on very large wheels.

Y, Z—Built-up gear wheel; another method for fastening
arms to boss on very large wheels.

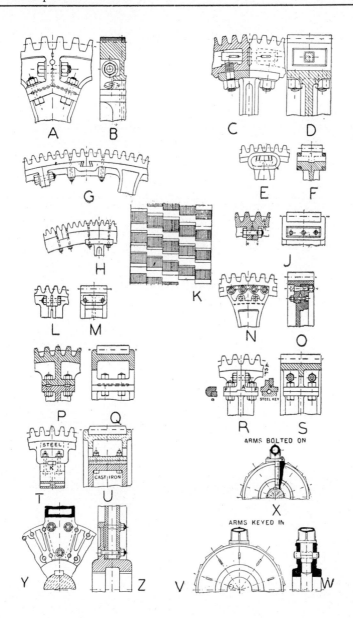

ARMS BOLTED ON

X

ARMS KEYED IN

V   W

CLASS IV.   BASIC MECHANICAL MOVEMENTS

Section 25a.   Toothed Gearing

A—Spur gears; parallel axes; external contact.

B—Spur gears; parallel axes; internal contact.

C—Spur gears; parallel axes; rack and pinion.

D—Spur gears; parallel axes; helical or twisted.

E—Spur gears; parallel axes; herringbone.

F—Spur gears; parallel axes; pin.

G—Bevel gears; intersecting axes; mitre bevel.

H—Bevel gears; intersecting axes; spiral bevel.

J—Bevel gears; intersecting axes; crown bevel.

K—Spiral gears; nonintersecting and nonparallel axes.

L—Worm and wheel; nonintersecting and nonparallel axes.

M—Hypoid gear; nonintersecting and nonparallel axes.

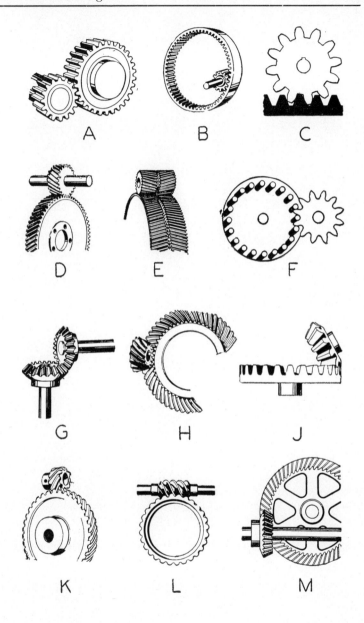

A

B

C

D

E

F

G

H

J

K

L

M

CLASS IV. BASIC MECHANICAL MOVEMENTS

Section 25b. Rack and Pinion

A—Ordinary rack and pinion for deriving reciprocating motion from circular or rectilinear.

B—Pump movement.

C—Circular rack and pinion gear.

D—Reciprocating motion of two pinions geared together.

E—Crank substitute; two loose pinions with reverse ratchets attached to a shaft with pawls on the pinion ratchets; each rack meshes with the reverse pinion for continuous motion of the shaft.

F—Rack motion for air pumps; the racks are directly connected to the pistons of a single-acting pump.

G—Reciprocating rectilinear motion of a double rack; it gives continuous rotary motion to the central crank; each stroke of the rack alternates on one or the other of the sectors; a curved stop on the center gear is caught on the pins in the rack to throw it into mesh with the opposite sector.

H—Sawmill feed.

J—Quick back motion.

K—Rectilinear vibrating motion.

L—Crank substitute for Parsons patent; a reciprocating double rack alternately meshes in a pinion; a cam face plate, running in smooth ways in the racks and fast to the pinion, lifts the racks into and out of gear alternately at the end of each stroke.

M—Doubling the length of a crank stroke.

N—Reciprocating rectilinear motion of a bar carrying an endless mangle rack; the pinion shaft moves up and down the slot, guiding the pinion around the end of the rack.

O—Alternate circular motion of a spur pinion from rectilinear motion of a mutilated rack gear.

P—Mangle rack; reciprocating motion of a frame to which is attached a pin-tooth rack, the pinion being guided by the shaft riding in a vertical slot, not shown.

Q—Vertical drop hammer movement or impact rod.

R—Mangle rack, guided by rollers and driven by a lantern half pinion. The long teeth in the rack act as guides to insure a tooth mesh at the end of each stroke.

S—Sector pinion and double rack; rectilinear reciprocating motion from the continuous rotation of the sector pinion.

T—Mangle rack with stationary pinion.

U—Alternate rectilinear motion from a swinging lever, with a quick return motion operated by a wrist pin on a face plate; the weight eliminates backlash.

V—Worm-screw rack.

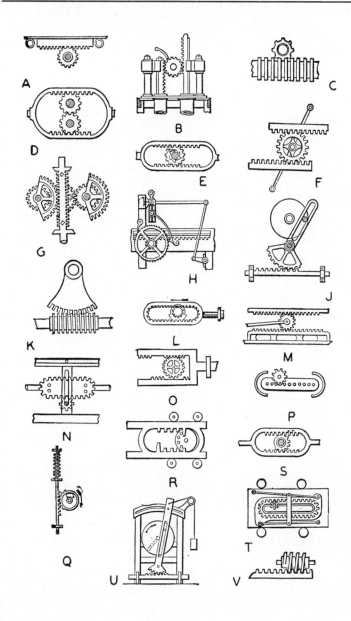

CLASS IV.   BASIC MECHANICAL MOVEMENTS

Section 25c.   Toothed Gearing

A—Alternate circular motion.

B—Mangle wheel with equal motion forward and return.

C—Mangle machine gear; the large wheel is toothed on both faces; the pinion moves from one side to the other through the open space.

D—Spiral-hoop gear for slow transmission to a shaft at right angles; one revolution of wheel A moves shaft B one tooth.

E—Continuous rotary motion of the pinion produces reciprocating motion of the double-geared wheel carrying the mangle drum which is given a quick return.

F—Mangle wheel that is given uniform motion through nearly a revolution and a quick return.

G—Rotary motion of a worm gear is given by the rotary motion of the wheel or vice versa.

H—Saw-tooth worm gear; it has a large area of contact.

J—Three-part worm screw for operating three screw gears for a chuck.

K—Right- and left-hand worm gear for operating drums or feed rolls.

L—Traversing motion; the sliding frame with the spur gear and worm is given a reciprocating motion equal to the throw of the crank pin.

M—Worm-gear pinion to drive an internal spur gear.

N—Release rotary motion; the weight D carries the loose arm quickly over a half turn, more or less as required; the worm wheel B lifts the arm and weight D beyond the vertical position by means of a pin on the shaft.

O—Antifriction worm gear; it has roller bearings on pins.

P—Release rotary motion; a sector weight E moves loosely on a shaft C to which a worm wheel B is fixed, driven by a screw.

Q—Release cam; uniform motion is given to gear wheel B fixed on its shaft with a pin C; the cam A is loose on the shaft with a stopped section to meet pin C. Lever D is raised by motion of the cam until its straight face reaches the roller, when it suddenly falls throwing the cam forward.

R—Alternating reciprocating motion, from two crank gears and walking beam.

S—Two-toothed pinion for transmission of motion to a wheel having teeth alternating on each side.

T—Pin wheel and slotted pinions.

U—Variable rotary motion from cone gears.

V—Scroll gear.

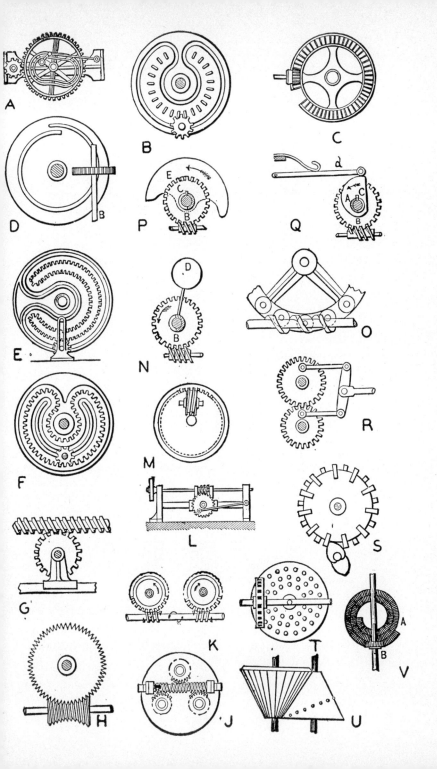

A

B

C

D

P

Q

E

N

O

F

M

R

G

L

S

H

K

J

T

U

V

CLASS IV.   BASIC MECHANICAL MOVEMENTS

Section 25d.   Toothed Gearing

A—Hunting-tooth worm gear; used for planetary or clock motion; the double-worm gear wheel may have one or several teeth more in one section than in the other; the motion of the worm advances one wheel in proportion to the difference in the number of teeth.

B—Complex alternating reciprocating motion from three unequal gears and two walking beams giving many varieties of motion to the connecting rod.

C—Accelerated circular motion.

D—Roller-bearing gear teeth.

E—Spiral gearing; silent transmission.

F—Expanding pulley.

G—Concentric differential speed; it consists of a high-speed shaft and an eccentric on which a low-speed gear A revolves with a differential motion by being carried around in mesh with the larger internal fixed gear C, giving a slow motion to belt pulley B.

H—Differential motions on concentric shafts by bevel gear.

J—Differential gear used in differential pulley blocks.

K—Variable-throw traversing bar; used in silk spooling.

L—Planetary motion.

M—Differential speed of two gears in different directions on the same shaft.

N—Capstan gear.

O—Slow forward and quick backward circular motion.

P—Geared grip tongs.

Q—Variable circular motion.

R—Alternating rectilinear motion.

S—Elliptical spur gear for variable speed.

T—Elliptical gear wheel and pinion for variable motion o¹ the pinion from the uniform speed of the elliptical gear

U—Irregular circular motion.

V—Intermittent motion of a spur gear.

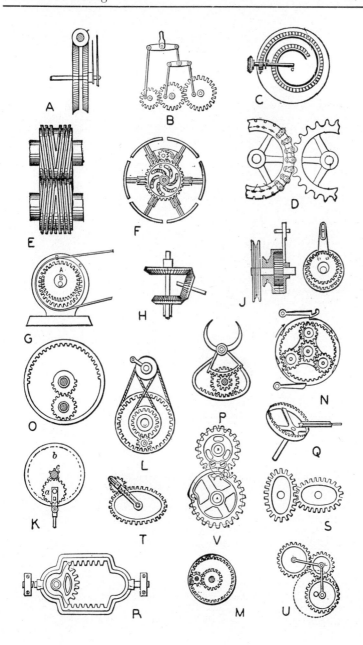

CLASS IV.   BASIC MECHANICAL MOVEMENTS

Section 25e.   Toothed Gearing

A—Variable reciprocating motion from a rotating spiral spur sector meshed in racks inclined to the plane of motion; the pitch lines of the rack are curved to fit the pitch line of the spiral sector; the pins on the sector mesh with the stop jaws *J, K,* on the rack frame, alternately at each half revolution.

B—Intermittent motion of a spur gear in which dogs *G* and *F* form a part of a driven gear *B;* this allows variable stop and speed motion of the two gears; *A* is the driving gear.

C—Spiral stop-motion gear; in addition to the stop, a variable motion is given to the driven wheel *B;* the dotted section at *G* shows the mesh of spur *K* of the stop wheel; *A* is the driving wheel.

D—Fast- and slow-motion spur gear; used also for quick return when operating a slide-crank motion.

E—Variable vibrating motion for rod *A.*

F—Motion by rolling contact of half-elliptical gears; a fork serves as a guide for the teeth.

G—Variable sectional motion from sector gears.

H—Miter intermittent gears; the driver makes one revolution to one-quarter revolution of the driven gears.

J—Uniform speed of a sectional spur gear during part of a revolution.

K—Intermittent rotary motion.

L—Scroll gearing for increasing or decreasing the speed gradually during one revolution.

M—Irregular vibratory motion of an arm *A* from the rotary motion of a pinion *B.*

N—Differential spur gear.

O, P—Stop-roller motion used in wool-carding machines; *P* shows the back of the disc *O.*

Q—Equalizing pulley for rope transmission; the arm carrying the smaller bevel gears is fast on a shaft; the divided pulley runs loose; any variation in the rope by tension will be compensated by the pinions.

R—Equalizing gear.

S—Change gear motion.

T—Double a revolution on the same shaft (Entwisle's patent); the pulley at *A* is driver on a shaft *D;* bevel gear wheel *A* is fixed; stud *E* is fast on the shaft; bevel gear wheel *B* revolves freely on stud *E;* bevel gear wheel *C* and its pulley *C'* run loose on the shaft; the rotation of stud *E* with its bevel gear wheel around the fixed bevel wheel *A* doubles the speed of the bevel wheel *C* and pulley *C'.*

U—Change gear motion; shafts *A* and *B* are disconnected and carry a loose hub and spur wheel in which is pivoted bevel pinion *T;* bevel wheel *C* is fast on shaft *A,* and bevel wheel *D* is fast on shaft *B;* any motion given to the central spur gear by the pinion shaft *E* varies the speed of the driven shaft *B* as compared to that of the driving shaft *A.*

V—Irregular circular motion.

W—Change gear motion with spur gears.

X—Ball gearing.

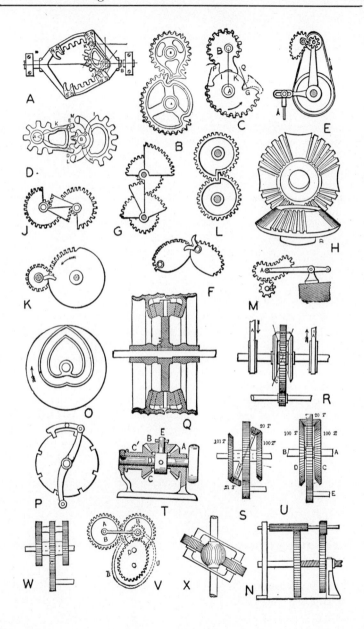

CLASS IV.   BASIC MECHANICAL MOVEMENTS

Section 25f.   Toothed Gearing

A—Epicyclic train; the gear wheel C is fixed and the arm D moves around its axis at A, the wheel B will have a retrograde motion and the wheel A a faster motion in the direction of the moving arm; if the wheel A is fixed, B and C will have unequal forward motions.

B—Sun and planet crank motion used by James Watt on the first steam engine; the wheel B is fixed to the connecting rod and does not revolve on its own axis, but moves around the axis of the fly wheel with a slightly oscillating motion; the wheel A revolves twice on its axis to one revolution of the wheel B, or to two strokes of the piston.

C—Epicyclic gear train.

D—Sun and planet winding gear.

E—Planetary motion as applied to an apple-paring machine.

F—Continuous shaft motion from an alternating driving shaft.

G—Epicyclic bevel gears; an arm FG is fast on a shaft AA; bevel wheel B is loose on this arm; bevel wheels D and C are loose on the shaft AA; differential motions of the two wheels C, D will produce rotation of the arm FG around and with shaft A, or, by making the arm loose on the shaft, a differential motion of the shaft and arm can be obtained.

H—Eccentric wheel train; an elliptical bevel gear A is fixed to the crank shaft, allowing bevel wheel B to clear bevel wheel F; bevel wheels B and D are fixed to a shaft H, giving to the shaft E an irregular reversed motion from the continuous motion at the crank shaft.

J—Alternating motion of a driven shaft at right angles to a driver shaft.

K—High-speed epicyclic train; bevel gear C is the driver.

L—Automatic clutch motion for reversing.

M—Planetary gear train; an arm T revolves around a fixed gear A on a stand H; gear B and bevel gear E are fixed on a shaft and turn in one direction, giving a contrary motion to bevel gear F and index pointer P.

N—Planetary gear train.

O—Change-motion gearing.

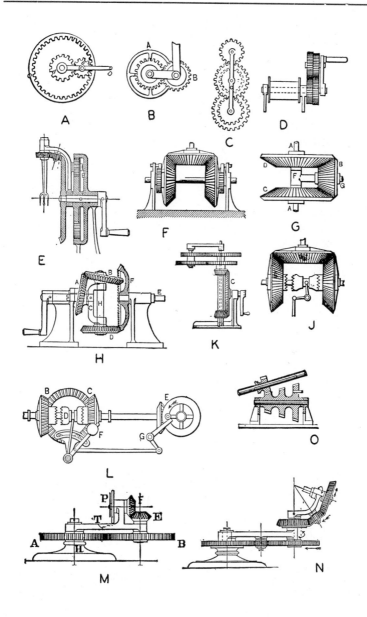

A

B

C

D

E

F

G

H

K

J

L

O

M

N

CLASS IV.   BASIC MECHANICAL MOVEMENTS

Section 25g.   Toothed Gearing

A—Spur wheels with long teeth or "star" wheels; used on roller mangles, where the centers rise and fall.

B—Pin wheel and pinion gear.

C—Lantern wheel.

D—Variable-speed square gear.

E—Irregular gear.

F—Moore's patent differential epicycloidal gear; the pinion and wheel are loose on the shaft and eccentric; one wheel has one tooth more than the other.

G—Differential gear; one wheel has one tooth more than the other.

H—Multiplying bevel gear; A is a fixed wheel; the cross C is keyed to the shaft; wheel B is loose on the shaft; bevel wheels D and E are loose on the cross C; the wheel B is driven at a speed greater than that of the shaft in proportion to the diameter of the gear.

J—Double worm gear with right- and left-hand threads; it neutralizes the end thrust on the shaft; wheels A and B may be geared together.

K—Ball miter gear.

L—Oval gear, linked together.

M—Wood-faced spur gear, to run quietly with the wooden faces in contact; the wooden faces are renewable like mortise teeth.

N—Elastic spur gear for preventing back lash.

O—Bevel gear with roller teeth in one wheel of the pair.

P—Plain bevel gear with shafts at obtuse angle.

Q—Conical rotary gear; used on reaping machines.

R—Forms of epicyclic or planet gear.

S—Scroll bevel gear.

T—Segment reversing gear.

U—Snail wheel or scroll ratchet.

V—Snail worm gear.

W—Worm and crown gear; obsolete mechanism for obtaining low speed on two shafts in opposite directions.

X—Combined spur and bevel wheel.

Y—Hook tooth and a pin gear.

Z—Ball wheel with limited angular traverse, gearing into one or two pinions.

AA—Skew bevels with the shaft off center.

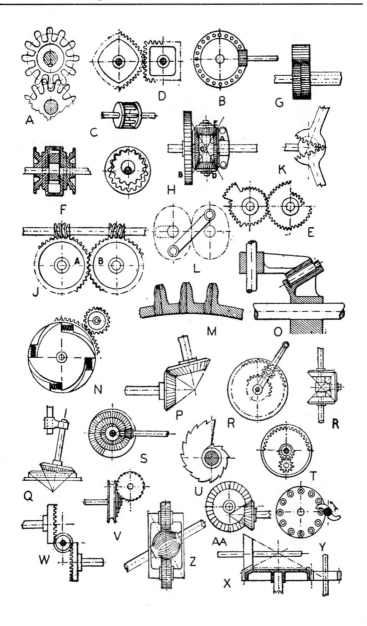

CLASS IV.   BASIC MECHANICAL MOVEMENTS

Section 25h.   Epicyclic Gearing*

A–Single planetary arrangement in which the planetary gear revolves within an internal stationary gear and around a central gear called the sun gear; the planetary gear rotates on its axis.

B–Compound planetary arrangement in which the sun gear is connected to the stationary internal gear by a compound planetary gear instead of a single planetary gear.

C–Planetary arrangement consisting of three planetary gears, mounted at the ends of three cams, forming a single unit called the spider.

D–Planetary arrangement consisting of four planetary gears mounted at the ends of four arms which form a spider.

E–Epicyclic train which may have any number of gears, some of which may be compound gears (two gears locked together or cut on the same piece of metal and turning as a unit), annular gears, bevel gears, etc.; if gear B has 100 teeth, gear C, 20 teeth and gear D, 60 teeth, one revolution counterclockwise of the arm A will cause gear D to make 7½ revolutions clockwise.

F–Epicyclic train in which gear K is a fixed internal gear; gears D and E are compound gears and rotate as a unit; all gears have the same circular pitch; if arm A makes 40 revolutions clockwise, gear B will make 30⅖ revolutions counterclockwise.

G–Epicyclic train; if arm A makes 20 turns clockwise and gear B 10 turns clockwise, gear D will make 30 turns clockwise and gear C, 13⅓ revolutions about its own axis; if arm A makes 10 turns clockwise and gear B, 20 turns clockwise, gear D will make 30 turns clockwise and gear C, 13⅓ revolutions about its own axis.

H–Section of automobile differential gear.

*From *Principles of the Basic Mechanisms* by courtesy of Department of Marine Engineering, U. S. Naval Academy, Annapolis.

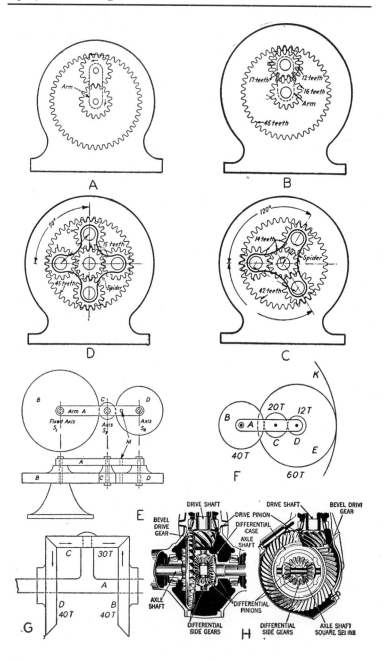

CLASS IV.   BASIC MECHANICAL MOVEMENTS

Section 26a.   Rolling Contact and Friction Gearing

A–Variable rotary motion from a friction pulley traversing a concave drum.

B–Variable motion to a right-angled shaft by curved conic friction pulleys with intermediate swinging pulley; used on light-power machinery, such as sewing machines.

C–Variable-speed gear used on "Wright's Model" sewing machines; the upper shaft is the driver, the lower shaft carrying the band pulley; the radius of frictional contact of the wheels varies as they are moved closer together or separated by the foot of the operator.

D–Robertson's patented wedge gearing with spring gripping adjustment.

E–Robertson's patented wedge gearing with weight and lever adjustment.

F–Jenkin's nest gearing; friction gear illustrating transmission of power between a shaft and dynamo; to obtain adjustments for the intermediate wheels D_1, D_2, D_3, the shafts are out of line, and the intermediate studs are fixed to a plate with curved slots.

G–Frictional rectilinear motion from an angular position of a sheave or pulley rolling on a revolving cylinder; A forward motion, B stop.

H–Friction-gear traversing motion.

J–Friction gear; variable speed from a pair of cone pulleys, one of which is the driver; a double-faced friction pinion moves on the line AB in contact with both cones.

K–Friction gear with variable speed.

L–Transmission of rotary motion to an oblique shaft by rolling contact of drums with concave faces.

M–Friction gear; a pair of friction discs on parallel shafts out of line, with a traverse pinion on a transverse spindle, will give a great range of speed.

N–Friction gear; variable-speed device for sewing machines; B is a swiveling yoke carrying a friction pulley and a driving pulley; the radius of frictional contact of the wheels varies as the surfaces are separated or moved closer together by radial motion of the yoke.

O–Friction gear (Howlitt's Patent); it consists of a rubber disc clamped between metal washers.

P–Grooved friction gearing

Q–Friction-gear variable motion to a shaft by curved-face discs, with a swinging rubber-tired pulley pivoted centrally to the curves on the face of the discs.

R, S–Wedge-surface friction gearing.

T–Plate friction clutch.

U–Magnetic clutch (Cutler-Hammer).

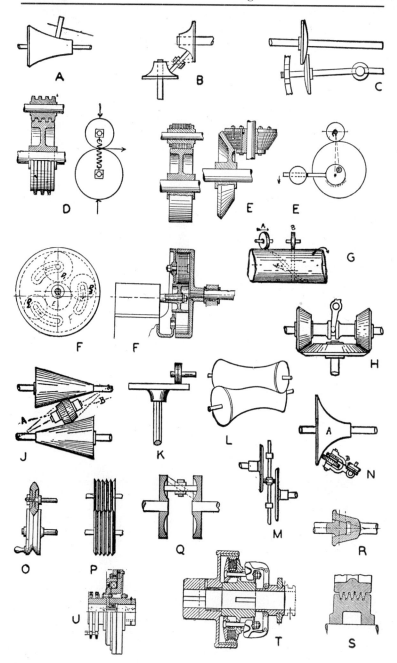

CLASS IV.   BASIC MECHANICAL MOVEMENTS

Section 26b.   Rolling Contact and Friction Gearing

A—Light friction drive by a solid wheel and rubber-tired pinion.

B—Friction spring clip for giving tension to cotton thread passed between convex discs.

C—Coupled bearings for friction gear to allow of any required pressure.

D—Disc wheel and rubber pinion arranged to reverse motion or vary speed; the motion is reversed by throwing either wheel into gear with the pinion; the speed is varied at will by raising or lowering the pinion; used for screw presses.

E—Common form of flat-faced friction gear for hoisting purposes; the required pressure is given by a weighted lever.

F—Frictional speed gear consisting of two hollowed discs revolving in opposite directions and geared together by one or more intermediate friction pulleys whose angle can be varied to give different speeds.

G—Carriage driving gear whose carriage wheel bears upward against the driving spindle which drives it by friction.

H—Variable-friction cone gear.

J—Variable-speed device consisting of two discs running in opposite directions on the same axis.

K—Variable-friction cone driving (Evans variable friction gear); a loose leather band, with a traversing motion by hand screw, forms the gripping medium between cones.

L—Variable-friction cup discs, with friction roller between them, whose angle can be varied; a variable-speed device.

M—Radial cones and endless grip belt, moving on a center, with guide sheaves carried in a radial frame.

N—Variable drive by a V-belt running between cone discs, the space between which can be varied by a hand lever or screw motion.

O—Centrifugal governor with cone-wheel motion to operate engine cut-off.

P—Antifriction rolling contact; the pivots of two rollers or shafts bear against the inside of a stiff ring producing rolling contact, but the rollers or shafts must run in the same direction; roller-mill application.

Q—Antifriction rolling contact; similar to P, but having three rollers or shafts.

R—Antifriction rolling contact; the shaft is guided vertically and its weight is carried by a larger roller with small pivots.

S—Antifriction rolling contact; the shaft runs in the V between two rollers.

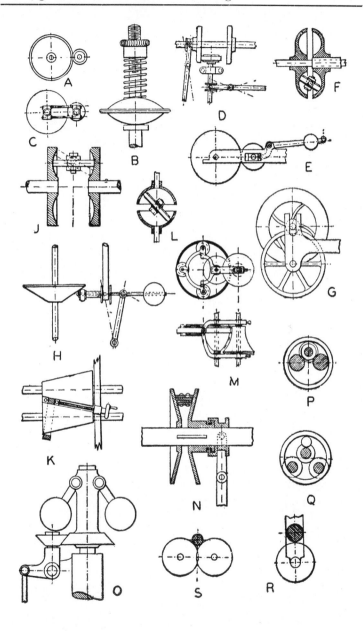

CLASS IV.   BASIC MECHANICAL MOVEMENTS

Section 26c.   Friction-Gear Details

A—Wedge gearing (patented by Robertson).

B—Wedge gearing of the continental type.

C—Friction gearing with two cylindrical wheels used to work a light windlass.

D—Simple rolling friction drive; two cast iron wheels A and B have their axes fixed, and the small wheel C faced with nonmetallic material, e.g., automobile brake lining is supported in such a way that it can be pressed against and withdrawn from A and B, as required.

E, F—Wood-faced pinions; the grain of the wood lies in tangential direction to the working surface, the different layers being fastened together with bolts, nails, glue, or white lead to form a friction driving wheel.

G—Pinion faced with leather, compressed-strawboard, or millboard (cemented under high pressure) is held between plates to form a friction driving wheel.

H—Disc or crown friction drive in its simplest form; disc B is pressed by force P against the edge of the other disc A, and as A is faced with softer material, it is, whenever practicable, made the driver; B, moving at a uniform speed, drives A nearer to the axis PQ, its speed decreasing gradually and being at rest when it reaches the axis PQ; if moved across the axis, its direction of rotation will be reversed.

J, K—Disc or crown friction drives.

L—Modified disc or crown friction drive; the two opposite discs B, B_2 (the drivers in this case) are forced by a lever against disc A, putting it into a condition of balance relative to the bearings AA and M, thus avoiding the wear o nbearings M and N in H, J, and K.

M—Conical or bevel friction wheels, used for high-speed turbines of small power; the iron wheel B is kept in frictional contact with bevel pinion A by spring CD acting on the end of the shaft of B, and the wheels are disengaged by pulling the bell-crank lever L into its dotted position L_2 which puts the spring out of action.

N—Wooden-face bevel pinion.

O—Leather or millboard bevel pinion.

P—Wooden-face bevel pinion.

Q—Leather-band friction wheel.

R—Variable-speed drive; a loose leather band with a traversing motion by a belt shifter A, B, supplies the gripping medium between the cones.

S—Disc, sphere, and cylinder drive (J. Thompson).

T—Engaging gear for friction drive (Lewis).

U—Bevel friction gearing.

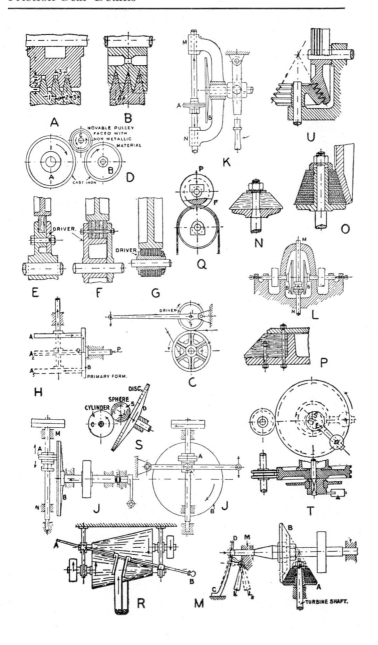

CLASS IV.   BASIC MECHANICAL MOVEMENTS

Section 27a.   Pawls and Ratchets

A—Intermittent circular motion from a vibrating arm and pawl acting on a ratchet wheel.

B—Ratchet lift; a vibrating lever operates two hooked pawls on the ratchet bar and lifts the bar; used in jacks and stump pullers.

C—Rotary motion from reciprocating motion of two racks meshing alternately with a gear wheel; the racks are pinioned at *aa;* the curved slots *bb* guide the racks into and out of gear; the bell-crank lever *c* and the spring *d* serve to disengage the rack at the end of the upstroke.

D—Double-acting click mechanism; the lever lifts the pawls, one of which moves the ratchet wheel at upstroke by one pawl, and again at downstroke by the other pawl.

E—Double-acting click mechanism; the vibration of lever *a* with its pawls *b, c* imparts a nearly continuous motion to the ratchet wheel; multiple-click mechanisms have three or more pawls; used where very large teeth are required, but the effect is the same as if there were more smaller teeth and only one pawl.

F—Intermittent rotary motion of a ratchet wheel by a lever and hook pawls.

G—Continuous feed of a ratchet by the reciprocating motion of a rod having two pawls on arms, and pivoted by links to the reciprocating rod.

H—Intermittent rotary motion; same as G, but acting on a face.

J—Ratchet-bar lift.

K—Pawl lift.

L—Ratchet head with spring pawls.

M—Ratchet intermittent motion by operation of treadles.

N—Intermittent circular motion from a reciprocating rod.

O—Intermittent circular motion; feed motion for planers.

P—Intermittent motion of a ratchet by oscillation of a knuckle-joint tappet arm.

Q—Oscillating motion into rotary motion by a straight and crossed belt running on two ratchet pulleys, the ratchets being keyed on the shaft.

R—Ratchet and lever pawl; the ratchet drops by weight of the lever; pulling of the rope unhooks the pawl.

S—Intermittent circular motion of a ratchet wheel with a check pawl by continuous circular motion of a pawl wheel.

T—Continuous rotary motion by check ratchet and oscillating beam.

U—Windlass grip pawl.

V—Ratchet governor for water wheels or other prime movers.

W—Intermittent rotary motion by ratchet and springs.

X—Intermittent motion of a ratchet crown wheel.

Y—Internal compound cam to operate a number of radial slides for internal grip.

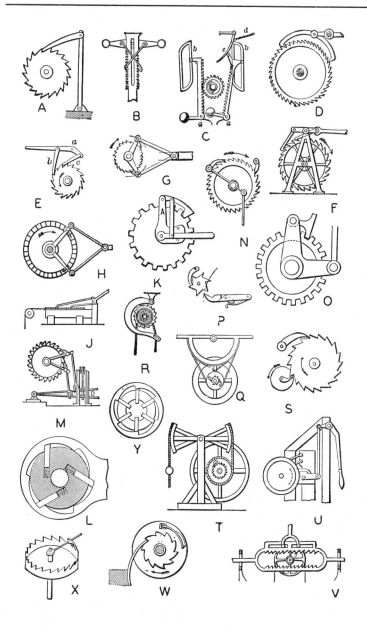

CLASS IV.   BASIC MECHANICAL MOVEMENTS

Section 27b.   Pawls and Ratchets

A–Intermittent circular motion from the oscillating motion of a lever by friction pawls.

B–Safety centrifugal hooks; centrifugal force throws the hooks out to catch pins on a plate.

C–Stops of various forms for a ratchet wheel; hook and straight-gravity pawl and a spring pawl.

D–Stops for a lantern wheel; one is a latch stop and the other a roller stop.

E–Stops for a spur gear; slip pawls.

F–Reverse ratchets, for continuous feed from an oscillating arm; it has two bevel gears and ratchets with pawls on opposite sides, so that there is a forward motion to the spindle at each stroke of the arm.

G–Ball and socket ratchet; the drill stock can be used at an angle.

H–Safety hooks for a hoisting drum.

J–Alternating rectilinear motion from studs on a rotating disc; the bar is carried forward by the stud on the disc striking the projection on the bar, and the bar returns by the movement of the bell-crank lever and opposite stud.

K–Vibrating toothed wheel for interrupting an electrical circuit.

L–Rack and pawl lifting jack; the lower pawl is operated by a lever.

M–Rocking escapement; the section teeth of the wheel pass the eye in the rocking cylinder at each quarter or at each half revolution.

N–Intermittent movement of a pin wheel by a hooked arm.

O–Intermittent motion of a toothed wheel.

P–Intermittent motion of a pin-toothed wheel by an indented tooth.

Q–Intermittent motion of a segmental-toothed wheel by the revolution of a segmental ring.

R–Elastic spur gear; it prevents back lash; the gear runs loose on the shaft; the ratchet wheel is fast on the shaft; compression springs are inserted between the shoulders of the gear and the cam ratchet wheel.

S, T, U, V, W–Intermittent rotary motions for counters and meters.

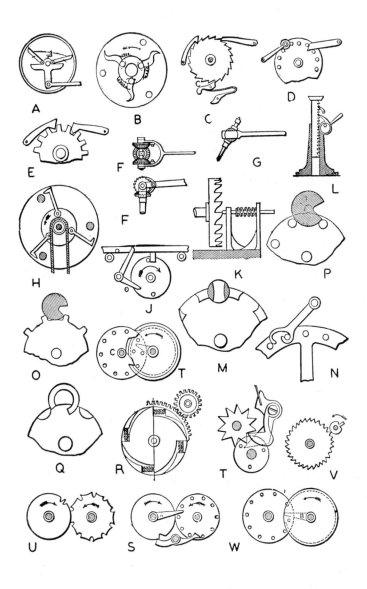

CLASS IV.   BASIC MECHANICAL MOVEMENTS

Section 27c.   Pawls and Ratchets

A—Strut-action pawl.

B—Rubber-ball pawl; sometimes a solid roller is substituted for the rubber ball.

C—Ratchet bosses.

D—Silent pawl; the pawl is lifted out of gear while reversing by motion of the lever and toggle joint.

E—Hare's foot ratchet motion with detent.

F—Reciprocating circular motion changed into intermittent circular motion (Kaiser's patent).

G—Continuous circular motion turned into intermittent circular motion; the wheel A is locked by ring C while finger B is out of gear; the ring then passes between the teeth of A (Kaiser's patent).

H—Slot wheel and pin gear.

J—Segment-wheel intermittent-feed motion; it is locked during the dead movement of the driving wheel.

K—Continuous circular motion turned into intermittent circular motion; the pawl is lifted out of gear at each revolution of the pin wheel and the ratchet moved one or more teeth for each revolution of the pin wheel.

L—Star wheel and fixed pawl; for conveying intermittent motion to a screw on a revolving disc; used for boring bars, slide rests, etc.

M—Pendulum and ratchet escapement.

N—Pendulum and double ratchet-wheel escapement.

O—Cylinder escapement.

P—V-pawl; it operates by wedging itself between V flanges.

Q—Gravity pawl and ratchet wheel.

R—Roller and inclined segmental recess for silent feed motion.

S—Friction-grip pawl applied to a wheel; it may also be applied to a rod.

T—Ratchet brace with slotted pawl.

U—Ratchet brace with friction-grip pawl.

V—Ratchet brace without pawl; the handle is hinged to the socket arm, and has a tooth gearing with the ratchet; it is thrown in and out by the movement of the handle.

W—Internal hooked pawl.

X—Internal strut-action pawl.

Y—Double-acting pawls and lever.

Z—Ratchet-rack, crank and connecting-rod intermittent movement; a detent may be added to return the rack.

AA—Gravity pawl and crown ratchet.

BB—Locked intermittent movement.

CC—Intermittent rotary movement on a spindle at right angles.

DD—Rack and screw press.

EE—Ratchet brace or feed lever in which the pawl is a fixed tooth and the lever is slotted to allow the pawl to clear the teeth on the back stroke.

FF—Silent feed; the jaw grips the rim of wheel when moving in one direction and runs loose in the other direction.

GG—Enlarged plan of cylinder escapement.

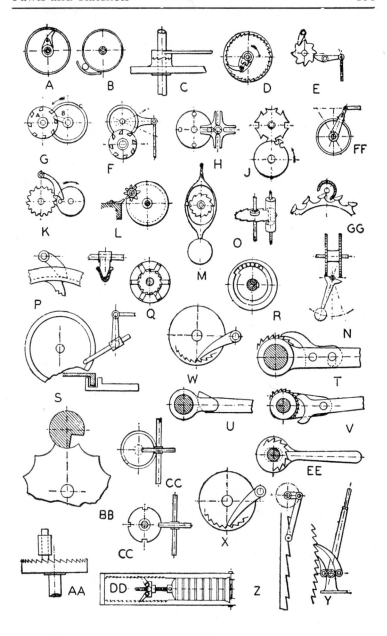

CLASS IV.   BASIC MECHANICAL MOVEMENTS

Section 28a.   Clock and Watch Movements

A—Centrifugal pendulum; the weight is driven in a circle by the clock movement.

B—Cycloidal pendulum movement.

C—Compound compensation pendulum; the upper part of the arms is made of steel which has a low coefficient of expansion and their lower part is made of brass which has a higher coefficient of expansion; by differential expansion of parts, an increase in temperature raises the weights to compensate for lengthening of the pendulum rod, and vice versa.

D—Compensating pendulum weight; a glass jar of mercury is used for weight, and is adjusted for length of pendulum by a screw and locked in place by a cross piece and catch; expansion by heat of the pendulum downward is balanced by expansion of the mercury upward and vice versa.

E—Compensating watch balance.

F—Anchor escapement in clocks.

G—Star-wheel escapement.

H—Crown-tooth escapement with ball balance.

J—Antique-clock escapement.

K—Lantern-wheel escapement.

L—Recoil escapement.

M—Stud escapement used in large clocks.

N—Pin-wheel escapement with a dead-beat stop motion; used for short-beat pendulum clocks.

O—Pendulum escapement.

P—Three-toothed escapement; a nearly dead-beat movement with long teeth and stops on the pendulum frame; A and B are pallets; E and D are stops.

Q—Hook-tooth escapement.

R—Detached pendulum escapement; in this movement, the pendulum is detached from the escapement, except at the moment of receiving the impulse from the single pallet.

S—Single-pin pendulum escapement.

T—Three-tooth pendulum escapement; impulse is given to the pendulum by contact of the pins with the pallets A and B alternately; the stops D and E hold the escapement during the extreme part of the pendulum stroke; the escapement makes one rotation every third stroke of the pendulum; the fly softens the strike of the pins on the pallets.

U—Three-toothed escapement.

V—Harrison winding device for clocks.

W—Mudge gravity escapement.

X—Bloxams gravity escapement.

Y—Dead-beat clock escapement.

Z—Endless cord winding device.

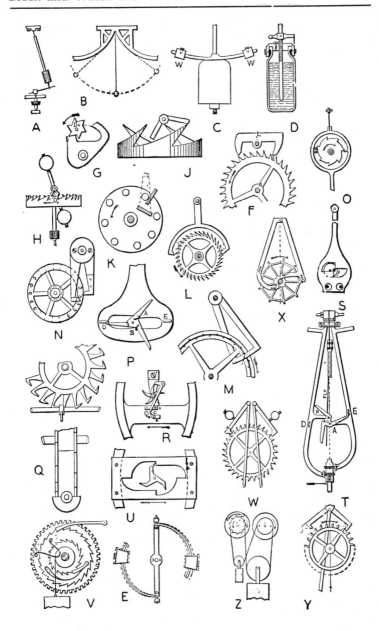

CLASS IV.   BASIC MECHANICAL MOVEMENTS

Section 28b.   Clock and Watch Movements

A—Minute-hand motion in a chronograph (A. Baud, Geneva).

B—Antique watch escapement.

C—Duplex escapement.

D—Verge escapement.

E—Watch regulator.

F—Guernsey escapement.

G—Lever escapement.

H—Arnold chronometer escapement.

J—Anchor and lever escapement for watches (Reed's patent).

K—Lever chronometer escapement.

L—Fusee chain and spring drum used in clock movements.

M—Geneva stop; a winding-up stop used on watches; it winds as many turns of the wheel A as there are notches in the wheel B, less one; the curve $a\ b$ is the stop.

N—Pin-geared watch stop.

O—Chronometer escapement.

P—Watch-stop.

Q—Watch train; a, key stem; b, barrel and spring and spur wheel; c, e, g, i, pinions; d, b, spur wheels; l, l, pallets and escapement; and k, lever and balance wheel.

R—Geared stop watch; contact of the two arms makes the stop.

S—Stem-winding movement of a watch; the movement of the lever with an arm outside of the rim locks a clutch on the hand gear; the third arm of the lever is thrown beyond the rim to prevent closing the case until the clutch is unlocked; it was used in the gay nineties.

T—Double tree-tooth pendulum escapement.

U—Clock train.

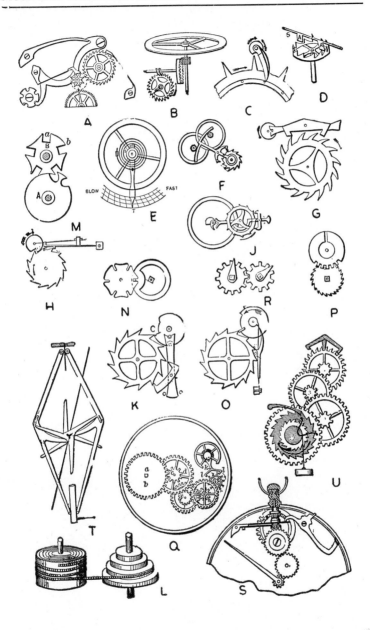

CLASS IV.   BASIC MECHANICAL MOVEMENTS

Section 29a.   Throwing in and out of Gear

A—The driving wheel is loose on the shaft and is locked to it by the hand-wheel nut, or by a ratchet wheel and locking pawl.

B—Radius bar and slot; the wheel can be shifted in or out of gear along the slot.

C—One shaft runs in eccentric bearings, which can be revolved so as to throw it out of gear with the other shaft.

D—Sliding-back shaft, which slides out of gear.

E—Bolt and slot device for gearing two wheels together on one shaft; used on lathe headstocks.

F—Worm gear which may be thrown out or in by moving the gear sideways on its shaft.

G—Half nut for throwing out of gear, with a screw and a spring to take up the wear of the nut.

H—Revolving worm for operating a belt-shifting bar and locking it at the same time.

J—Belt-shifting bar, adjustable in every position.

K—Motion for punching machines; it is used to set the punch in or out of action by a cam and hand lever.

L—Method of throwing a pulley out of gear by slacking the belt; this is done either by a cam bearing or a sliding motion to the driven shaft; it works best in the vertical position.

M—Cam slot motion for a back shaft to throw it in or out of gear.

N—Presser foot for sewing machines or for the intermittent holding of any flat article; lifted out of gear and held by the keyer and resting on the sliding socket.

O—Two half nuts are lifted in or out of gear with a screw by cam or lever action.

P—Sliding shaft for winch or other gear to shift a pinion out of gear or change to another speed.

Q—Device for locking a sliding shaft in or out of gear.

R, S, T—Conventional gear and wheel pullers.

U—Chain-gear puller invented by the author (licensee Delavan Mfg. Co., Des Moines, Iowa).

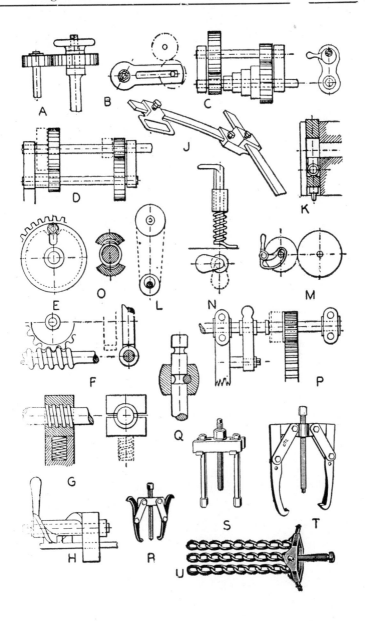

CLASS IV.   BASIC MECHANICAL MOVEMENTS

Section 29b.   Miscellaneous Shifting and Reversing Gears

A—Leather-belt shifter; when the belt is shifted from the loose pulley L to the tight pulley F, the pulley C revolves and operates a lathe; to stop, the belt is shifted back to the loose pulley L by pulling D.

B—Belt double-shifting device; shifting to left gives one speed and shifting to right gives another speed; when the lever is on dead center, both belts are on loose pulleys L.

C—Stepped-cone gear with four speeds, showing belt tightener a.

D—Reversing pulley.

E—Four-speed change gear; a hollow spindle with change gears running loose on it; rack spindle B carries a hinged pawl or key A, held out by a spring; a lever C carries a sector meshing in the race, which by its movement draws the key A to catch the keyway in any of the speed gears.

F—Gear shifter.

G—Shifter for throwing a friction clutch in and out of gear.

H—Reversing gear from a single belt and cone pulley.

J—Reversing lever with rack sector and worm gear; the worm wheel is lifted from the sector for large movements by the small latch lift and snaps back while a small movement is made by the handle at the top of the lever.

K—Reversing movement for a pump valve; the piston-rod trip carries the ball frame beyond the level, when the ball rolls across and completes the value throw (a very old device).

L—Stop, driving and reversing motion with a single belt, which may be operated from the drum on the driving shaft or from a bevel gear on shaft C; middle pulley being loose on shaft a, right-hand pulley tight on shaft a, left-hand pulley tight on hollow shaft Bb; the operation of a single shifter changes the motions or stops.

M, N—Manual belt-reversing shifting gear.

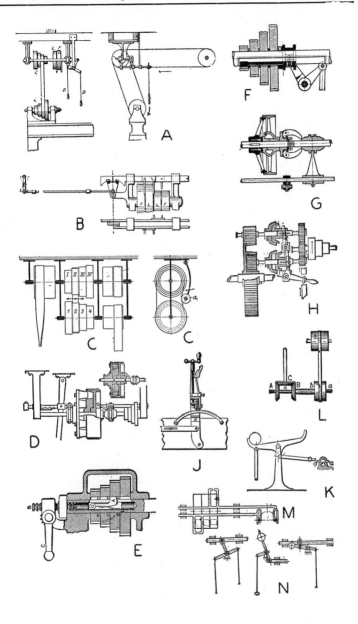

CLASS IV.   BASIC MECHANICAL MOVEMENTS

Section 29c.   Hinges

A—Common double-leaf hinge.

B—Elevator hinge; causes door to rise slightly on opening; it will then close by means of its own weight.

C—Cup and ball hinge.

D—Pintle hinge.

E—Hinge for a door that lies flat against a wall at either side when open.

F—Door spring hinge; it returns the door to central position; the cams press against a roller attached to the springs.

G—Door hinge with tension springs; similar to F.

H—Door spring hinge with tuggle movement.

J—Gate hinges; it has a double pintle at the bottom to center the gate when closed.

K—Link hinge for trap door or grid to allow it to lie flat when opened.

L—Multiple hinges with one bolt; for long or heavy doors.

M—Hinge permitting the door to lie flat against a wall.

N—Link hinges used for reversing car seats.

O—Tape hinging to allow a door to lie flat against a wall at either side when open.

P—Link hinges for reversing a shutter or door.

Q—Door hinged to swing through 360°.

R, S—Spring hinges.

T—Hinged handle for combined latch and staple.

U, V, W—Hinges for heavy doors.

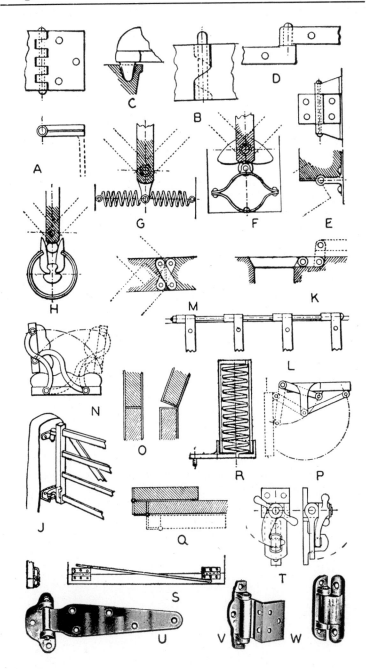

CLASS IV.   BASIC MECHANICAL MOVEMENTS

Section 29d.   Centers and Spindles

A—Plain spindle with two loose collars.

B—Spindle with one collar and one sunk bearing.

C—Spindle with sunk end bearings.

D—Fixed coned center.

E—Fixed-collar center pin or stud bolt.

F—Coned center; drive fit.

G—Square center.

H—Parallel center for rollers, etc., keyed in place.

J—Lathe fast headstock spindle.

K, L—Lathe headstock spindles.

M—Conical crane post.

N—Special car axle.

O—Universal centers.

P—Square-neck center bolt.

Q—Coned and cottered crank pin.

R—Center pin and bracket.

S, T—Two means of securing the end of a rod to a machine part.

U—Hollow post center with steam or water channel to permit swivelling.

V—Sleeve centers.

W—Plain center pin with split pin, nut and washer.

X—Plain center pin with washer and split pin.

Y—Hook center pin; it is readily disengaged.

Z—Stud center with riveted washer or nut.

AA--Center pin with screw and lug.

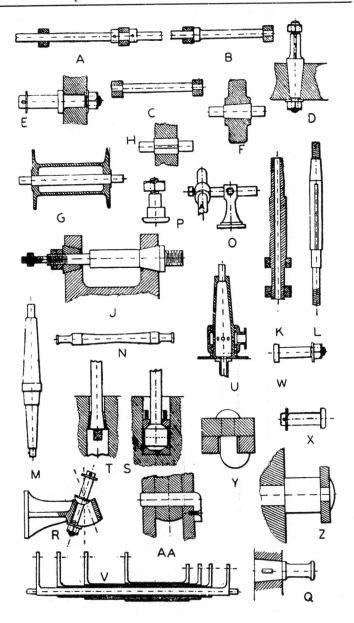

CLASS IV.   BASIC MECHANICAL MOVEMENTS

Section 29e.   Transmission of Motion

A, B, C, D, E, F—Trunk piston centers.

G, H, J—Iron centers for a wooden shaft.

K—Eyelet center.

L—Swaying ball center.

M, N—Swaying or rocking centers.

O—Swinging pipe joint.

P—Universal hinge; the arm can be fixed in any position.

Q—Ball joint; it can be fixed in any position.

R—Ball and cup joint.

S, T—Hooke's universal joints.

U—Universal hinge; it can be fixed in any position by tightening the gland.

V—Ball castor.

W—Typical line shaft.

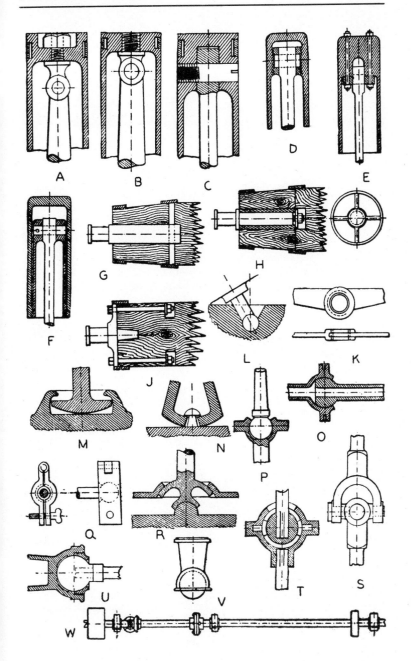

CLASS IV. BASIC MECHANICAL MOVEMENTS

Section 30a. Circular and Reciprocating Motion

A–Variable reciprocating motion.

B–Trammel gear; the slotted crosshead moves at right angles across the path in the cross slot.

C–Slotted lever motion.

D–Intermittent reciprocating motion.

E–Variable crank throw.

F–Variable adjustment for spring tension on motion of a connecting rod by changing the radii of a rocking lever.

G–Equalizing tension spring and lever.

H–Variable crank motion; an eccentric slot in a stationary face plate guides a slide block and wrist pin in a slotted crank; the connecting rod drives the cutter bar of a planing machine.

J–Variable rectilinear motion; an oblique disc drives a rod.

K–Traverse mechanism and grooved cam.

L–Ovoid curve made by any point between the crosshead pin and the crank pin.

M–Variable power transmitted from a crank linked to a lever beam driving a second crank; there is no pressure on the driven crank when both cranks are vertical, but the pressure increases as the position of the crank changes and will be the greatest when the cranks are horizontal.

N–Spring lathe-wheel crank; spring *A* is intended to keep crank *B* off the dead center.

O–Reciprocating motion of a connecting rod through a bell crank connected directly with a wrist on the crank disc.

P–Variable circular motion from two cranks on parallel shafts, but out of line.

Q–Irregular motion of one crank from the regular motion of another crank.

R–Vibrating motion. A slotted curved arm gives a variable vibrating motion to a straight arm.

S–Variable crank throw.

T–Variable-radius lever.

U–Variable crank throw by a movable pin block in a slotted face plate and transverse screw.

V–Rotating slotted crank.

W–Rocking motion.

X–Multiple-return cylinder.

Y–Slotted-yoke crank.

Z–Elliptical crank-end motion.

AA–Bobbin winder; the flyer revolves, while the bobbin is moving up and down the spindle for even winding.

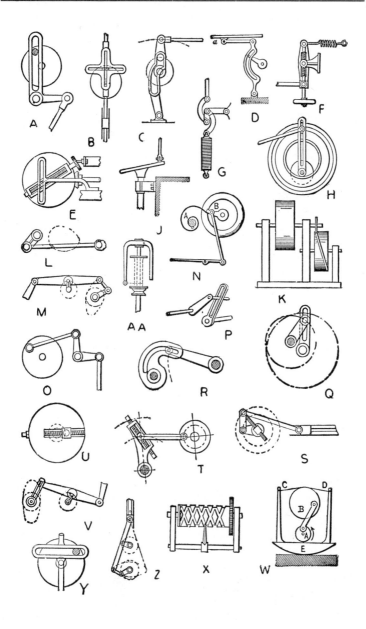

CLASS IV.   BASIC MECHANICAL MOVEMENTS

Section 30b.   Circular and Reciprocating Motion

A—Ordinary piston-rod and crank motion.

B—Patent crank motion (Bernay); the radius of the crank is $\frac{1}{4}$ of a stroke.

C—Friction gear; it is loose on stroke to left and gives friction grip on stroke to right.

D—Lever and roller crank pin.

E—Treadle motion.

F—Oscillating clutch arm and ring; silent-feed motion.

G—Reciprocating motion derived from circular motion.

H—Slot link and treadle, driving the pinion by friction on the inside of the link.

J—Double crossheads; separated to permit one connecting rod to work between them.

K—Suspended treadle action.

L—Rocking-lever motion.

M—Crank pin and slotted lever for giving a variable speed to the connecting rod.

N—Crank motion to work a sliding tool or movement on a bar or guide.

O—Bell-crank and disc-crank motion.

P—Worm wheel and screw reciprocating motion by means of a tied crank pin; useful for low speeds.

Q—Reciprocating motion derived from circular motion or vice versa.

R—Bent-shaft and arm motion.

S—The "Dake" square-piston engine.

T—Crank motion; the crank pin runs in a sleeve sliding along a lever.

U—Double-piston crank motion with side connecting rod and off guide.

V—Ball and socket crank motion.

W—Side-gudgeon crank motion.

X—Crank motion without dead center; it has a slotted crosshead.

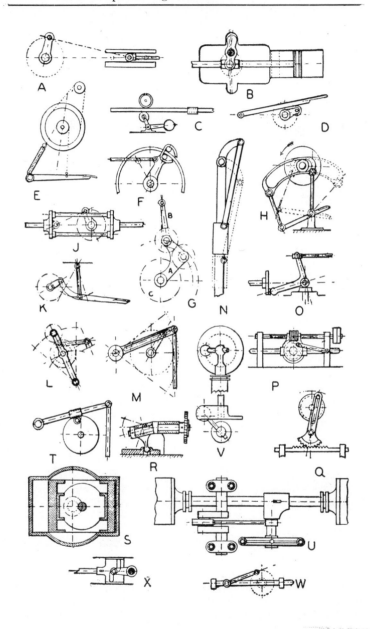

CLASS IV.   BASIC MECHANICAL MOVEMENTS

Section 30c.   Circular and Reciprocating Motion

A—Double-piston crank motion.

B—Crank motion similar to A, but with yoke connecting rod.

C—Double-piston crank motion for pump or compressor; its lever describes an elliptic arc.

D—Bouchet's crank motion to avoid dead center.

E—Circular motion converted into reciprocating motion by diagonal grooved sheave; the crank arm center is in line with the center of the sheave (J. Warwick's patent).

F—Double-geared cranks.

G—Trammel gear with one revolution of the wheel to two double strokes of the piston.

H—Reciprocating wheel and crank motion.

J—Eccentric and sliding-bush motion for double-piston engine.

K—Offset crosshead and side-crank motion for an air compressor or pump.

L—Eccentric and crank motion; the connecting rod has a ring-shaped end and the strap is revolved on the center pin by a lever fixed to it.

M—Atkinson's crank motion which drives a fly wheel two revolutions to one double stroke of the piston.

N—Crank motion to drive a swing arm or swing arm to drive a crank.

O—Slide crank motion.

P—Treadle motion without a dead center; spring A forces the crank past the centers.

Q—Reciprocating motion of a rod moving in a guide.

R—Reciprocating and rotary motion with pause at each end of the stroke.

S—Knight's noiseless gearing for two shafts running in opposite directions; each shaft has two equal cranks at right angles, which are coupled by links to rocking arms, which are also coupled in pairs.

T—Intermittent reversible feed motion; the pinion is of rawhide and drives the segment until it runs out of gear.

U—Double-screw arrangement for a steering gear.

V—Spring friction-grip wheels; the crank may be placed on the disc.

W—Annular ball-jointed crank motion.

X—Angle coupling (Dr. Hooke's principle).

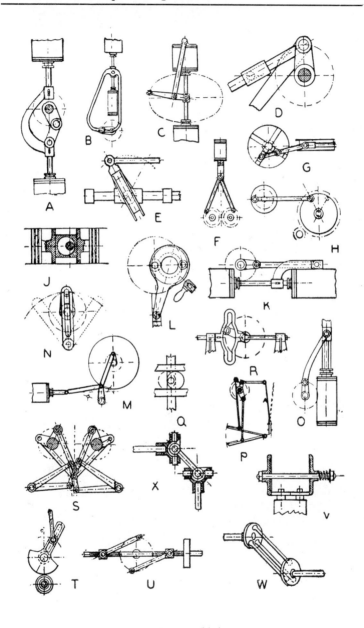

CLASS IV.   BASIC MECHANICAL MOVEMENTS

Section 31a.   Variable Motion, Speed and Power

A—Eccentric gearing; wheel *A* is fixed on the crank pin; driving wheel *B* drives the dotted gear at a speed proportional to the diameters of wheel *A* and the driven wheel.

B—Multiple trammel gear; the pinion is half the diameter of the wheel and makes two revolutions to one of the wheel.

C—Trammel gear; the crank revolves once to two double strokes of the piston.

D—Variable crank pin adjusted by a sector and bolt.

E—Eccentric variable-speed toothed gear.

F—Beam motion with variable fulcrum for changing the lengths of stroke of the driving and driven cylinders.

G—Variable-pressure accumulator; both cylinders are connected by a pipe, and the pressure varies with the angle of the ram.

H—Convex and concave cones for belts.

J—Three-speed gear; each separate pair of spur gears is driven by its own belt on separate sleeves.

K—Owen's compound-lever variable-perssure air pumps; the pressure increases and the speed decreases as the pistons reach the top of their stroke.

L—Variable-fulcrum lever with shifting pin and hole adjustment.

M—Variable travel of a piston rod from the crank by changing the point of attachment of link *A* to the slot.

N—Variable-throw crank pin by means of a jointed crank and radial adjusting screw.

O—Similar to L but operated by shifting the fulcrum point along the slot.

P—Variable-throw crank pin; it is connected to an eccentric disk.

Q—Variable belt drive by elliptic pulley.

R—Variable-power pistons; single acting.

S—Adjustable center-piece, or bearing for a spindle or rod.

T—Cone change-speed device and endless belt.

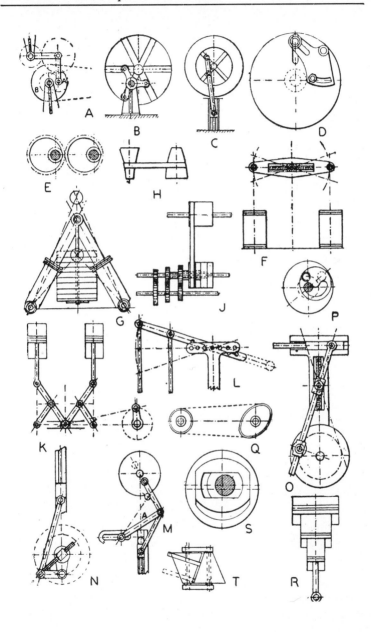

CLASS IV.   BASIC MECHANICAL MOVEMENTS

Section 31b.   Variable Motion, Speed and Power

A–Variable-driving friction gear.

B–Variable-throw crank pin.

C–Variable-radius head crank.

D–Variable-diameter belt pulley having six small pulleys.

E–Expanding V-belt pulley.

F–Belt speed cone device.

G–Belt cones and endless belt; the belt is kept flat by two endless bands tapered to the angle of the cones, having flanges to prevent the driving belt from running off.

H–Pair of variable-diameter belt drums.

J–Newman's variable-speed device; it consists of a variable-throw eccentric, the strap of which is connected by four links to four clutches and pinions; the latter are geared to a central wheel; the movement of the eccentric strap is conveyed in revolving to the clutches and pinions, and by them drives continuously the central wheel at a speed depending on the eccentricity of the eccentric.

K–Four-speed spur gear; any pair of the wheels can be coupled by a sliding key on one shaft, the others running loose.

L–Four-speed gear with three shafts and three clutches, *A* or *B* being the driver.

M–Three-speed gear with separate clutch for each pair of wheels.

N–Four-speed gear; *A, B* and *C* slide together on a key on the shaft.

O–Four-speed face gear.

P–Six-speed bevel gear; the pinion on the diagonal shaft can be slid into gear with any of the six bevel wheels.

Q–Four-speed spur gear with a sliding shaft and a short key.

R–Two-speed bevel gear with three wheels and a sliding shaft, by which two of the wheels can be put into gear.

S–Two-speed bevel gear with four wheels and a sliding shaft.

T–Irregular or elliptical spur gear.

U–Scroll gear for obtaining a variable pull from a weight.

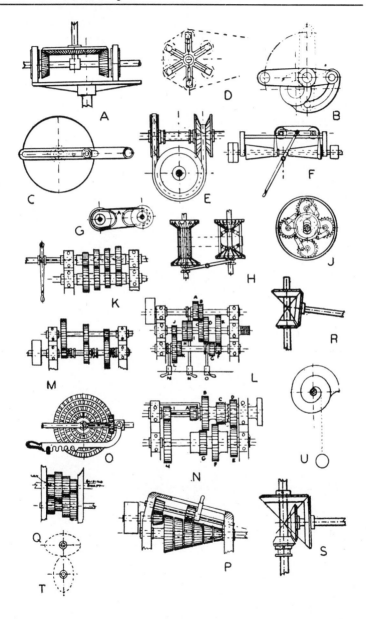

CLASS IV.   BASIC MECHANICAL MOVEMENTS

Section 32a.   Differential Gearing

A—Equational box; two drivers A, A', equally speeded in opposite directions, will drive the bevel gear at the same velocity without revolving the spur wheel C' which is loose on the shaft; but any alteration in the relative speeds of A and A' causes the bevel pinion to travel round, carrying the spur wheel C' at a speed equal to half the difference of the two velocities.

B—Modification of equational box; the pinion A may be controlled in speed by any manual or automatic device to vary the speed of the driving pinion B. The pulley C carries round the bevel wheel D at a speed varying with the motion given to A.

C—Two wheels, one of which has a number of teeth different from the other, mashing into one pinion.

D—An application of C by internal or epicycloidal gear to pulley blocks.

E—Weston's differential pulley block consisting of a two-grooved, pitched chain-sheave having different numbers of teeth, in combination with an endless chain and return block.

F—Differential screws; these may be both of the same hand, or one right-handed and one left-handed; any fractional speed can be secured by proportioning the pitches.

G—Stewart's differential gear.

H—Two-speed gear, operated by a double clutch, which throws either pair of wheels into gear as required.

J—Differential hydraulic accumulator; the effective area of the ram is the annular shoulder, or the difference between the areas of the top and the bottom rams.

K—Differential governing device; the motive power drives pulley A which winds up the large weight; the small weight tending to run down, drives the fan regulator, and the two weights are so adjusted that when the proper speed is obtained, both weights are stationary; any change of speed causes them to run up or down, so actuating the regulation by the bell-crank lever and rod.

L—Varying differential regulator; the upper rod A is connected to the regulator valve or other device, and it is capable of receiving motion from either the piston, which acts against a spring, or from rod B attached to same positive reciprocating attachment, so that the full movement of A is due to the difference between the motion of rod B and piston C.

M—Differential-piston indicator for steam engines.

N—Harrison's differential epicycloidal hoist gear; pinion A is fast to the barrel and loose on the shaft, B is keyed to the shaft, C and D are cast together, and run on a stud in the large wheel E, which is loose on the shaft; A and B have different numbers of teeth.

O—Differential-screw bolt and sleeve movement.

P—Differential screw-valve fitting, with cone seat, tightened by the T-head and fine-thread central screw; used for gas containers.

Q—Chinese windlass; ancestor of modern differential gear.

R—Differential drive for swivelling wheels; F and G are long pinions geared together; F meshes with A and G meshes with B, which is fixed to C, and both are keyed to the shaft; A drives one road wheel by its sleeve and B the other, the road wheels being not shown.

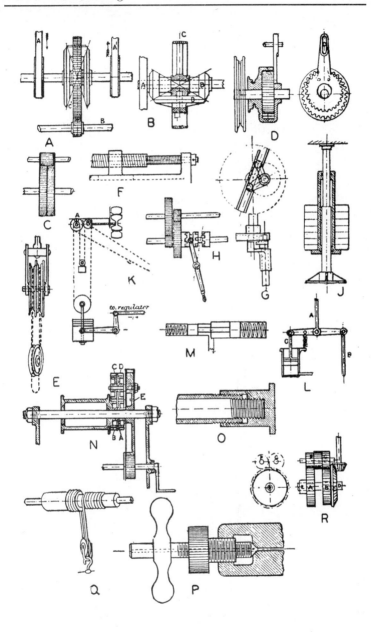

CLASS IV.   BASIC MECHANICAL MOVEMENTS

Section 33a.   Augmentation of Power

A—Compound lever.

B—Double toothed cam and lever combination.

C—Double lever and link motion with increasing pressure.

D—Lever and toggle motion.

E—Compound-lever shears.

F—Lever and frame gear for great multiplying leverage, and detent to prevent slipping back.

G—Compound-lever cutting shears.

H—Stake puller.

J—Hydraulic press.

K—Portable riveter; the large piston and lever give great power.

L—Lever toggle joint used in stamping presses.

M—Screw stamping press.

N—Double-screw toggle press.

O—Single-toggle-arm letter press.

P—Toggle-joint stone breaker.

Q—Lewis' wedge for lifting stone.

R—Toggle-joint stone breaker.

S—Weston differential gear hoist.

T—Toggle-bar press; the toggle bars have spherical ends.

U—Sector press.

V—Adjustable grip tongs.

W—Lever grip tongs.

X—Compound tire-upsetting and punching machine.

Y—Screw jack.

Z—Letter press with Stanhope levers.

AA—Toggle-joint wagon brake.

BB—Rack and lever jack.

CC—Hydraulic jack.

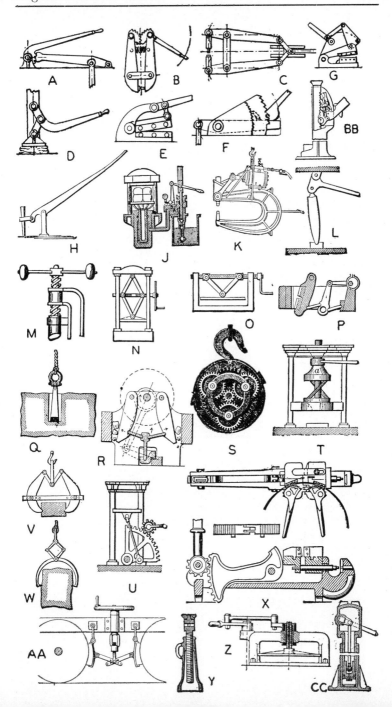

CLASS IV.   BASIC MECHANICAL MOVEMENTS

Section 34a.   Transmission of Motion

A—Flexible shaft for light driving; used for drilling in difficult positions.

B—Radiating arm and bevel gear; the movable machine can be driven at any point within the circumference of the circle described by the arm head.

C—Bevel gear and feather shaft; the movable machine travels in a straight line in the length of the shaft and also radially.

D—Screw and worm-wheel gear; for the same purpose as C.

E—Steam or hydraulic radiating arm and cylinder device; for the same purpose as C.

F—Central-cylinder and radiating-lever motion for the same purpose as C.

G—Jointed radiating arms, with belt gear for conveying motion from a central spindle to one having a travel covering any point within a circle of the extreme radius of the jointed arms.

H—Jointed tube for a traveling, hydraulic, steam or compressed-air hoisting or other engine.

J—A traveling wheel may be driven by a long pinion without affecting the movement of the wheel.

K—Traveling spur gear similar to J to convey continuous motion to a traveling machine.

L—Telescopic swinging-shaft movement with universal joint at each end.

M—Helical-gear drive for a traveling pinion.

N—Endless cord drive to a machine having limited movement.

O—Parallel-motion radiating driving device; with a limited vertical travel and a radial motion.

P—Traveling gear and slot; the driven wheel *A* has a limited travel up and down the slot, the idle gear wheel *B* being kept in gear by the link suspension.

Q—Radiating arm and belt; the movable machine can be driven at any point in the circumference of the circle described by the arm head.

R—Endless, round rope-belt drive; it is kept tight in any position in the plane of the driving pulley by a weighted pulley; the machine can be moved to any position in the plane of the driving pulley, the weighted pulley taking up the slack in the belt.

S—Motion conveyed by belt to a driven shaft having a radial motion in a vertical plane.

T—Rock drill operated by compressed air, with hose connection.

U—Locomotive air brake.

V—Flexible steam joint.

W—Transmission of motion by electricity.

X—Electric and magnetic waves generated by an oscillating electric current.

Y—Electric hoist.

Z—Idle wheel and slot; a common device for changing direction or speed in driving gear by connecting or disconnecting it with intermediate gearing between a fixed driving and a driven shaft.

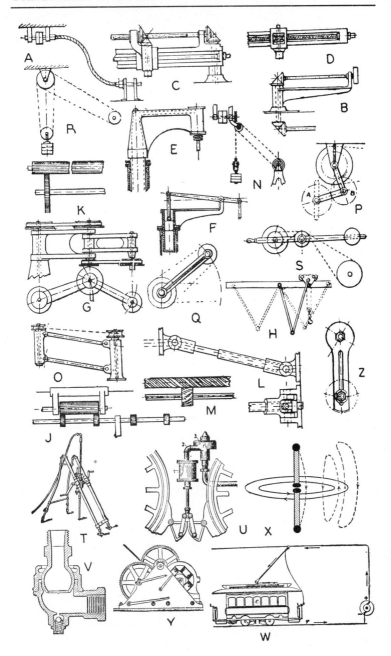

CLASS IV.   BASIC MECHANICAL MOVEMENTS

Section 34b.   Flat-Belt Transmission

A—Quarter-turn method of flat-belt arrangement with shafts at right angles.

B—Arrangement for shafts inclined to each other; the center line of the belt advancing on the pulley should be in a plane passing through the midsection of the pulley at right angles to the shaft.

C, D—Guide pulley positions for an inclined shaft.

E—Three-step cone pulley.

F—Five-step cone pulley.

G—Belt gear for variable-speed machinery.

H—Safety cap for gib-headed key.

J, K—Guide pulley positions for an inclined shaft.

L—Use of guide pulleys to lengthen the belt when the shafts are too close together.

M—Guide pulleys with their axes in line; pulleys M are moved to right to receive the belt from pulley A.

N—Use of guide pulleys with shafts at right angles.

O—Belt gear for slow forward and quick return motion.

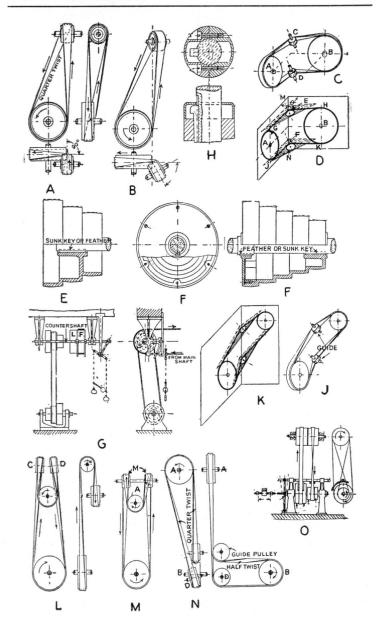

CLASS IV.   BASIC MECHANICAL MOVEMENTS

Section 34c.   Belt Tighteners and Pulleys

A—Belt tightener; adjustable by spring tension or weight.

B—Alternate method of belt tightening.

C—Tandem drive.

D—Flanged pulley.

E—To shift the position of a belt on a pulley, the advancing part must be displaced by the shifting fork.

F—Displacing the retreating part of a belt is wrong.

G—Crowned pulley; the belt will climb to the largest diameter and remain there; shifting or idler pulleys are flat while power transmitting pulleys are always crowned or rounded.

H—Vertical pulley; it has a flange or shroud at the bottom to take up the weight of the belt.

J—Fast and loose pulley arrangement for relief tension.

K—Fast and loose pulleys for the shaft of a machine.

L—Detail of a loose-pulley bushing.

M—Fast and loose pulleys without bushing.

N—Pulley arrangement on the driving shaft with relief tension.

O—Wall guide pulley and fittings.

P—Ceiling or floor guide pulleys.

Q, R, S, T, U—Miscellaneous flat-belt arrangements.

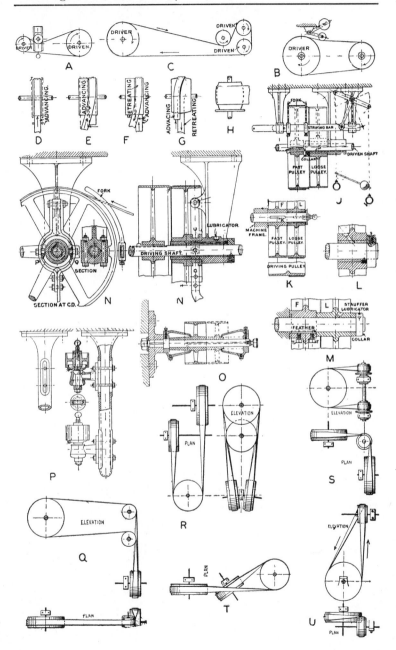

CLASS IV.   BASIC MECHANICAL MOVEMENTS

Section 34d.   Solid and Split Pulleys

A, B, C, D, E—Sections of pulley rims.

F—Pulley with straight arms.

G—Pulley with curved arms.

H—Section views of pulleys.

J—Pulley with double curved arms.

K—Wooden split pulley.

L, M—Cast-iron split pulleys.

N—Wrought-iron split pulley.

O, P, Q, R—Steel split pulleys.

S—Bed for countersunk head of rivet formed by forcing the rim of the pulley into the countersunk hole in the arm.

T—Medart pulley whose arms and boss are made of cast iron in one piece and the arms are riveted to a wrought-iron rim.

U—Expanding pulley.

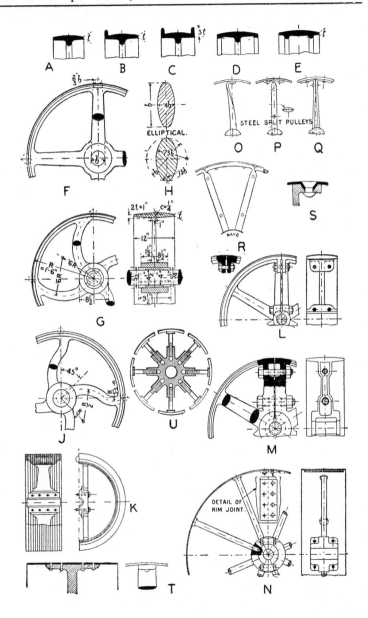

CLASS IV.   BASIC MECHANICAL MOVEMENTS

Section 34e.   Belt Joints and Fasteners

Belt Materials. Oak-tanned, center-stock, double-ply leather, made endless by cementing, is usually considered best for heavy-duty work. There are two general methods of arranging short-center drives of the flat-belt type. One consists in applying an idler pulley or tension roller on the slack side, near the smaller pulley. The other type consists of a pivoted motor to increase or decrease belt tension. Flat belts made of rubber are used in damp places. Textile belts made of cotton and hair are used where first costs are limited. When considered over a period of years, leather is more reliable than rubber or cotton flat belting. Thin flat steel belts have been used in Germany and the users claim many advantages.

V-Belt Drives. This type of belt provides a compact, resilient transmission and has been applied extensively to automotive drives for fans, generators, refrigerating units and industrial transmission. V-belts in sizes from fractional to 6,000 horse power constant-speed, and to 300 horse power adjustable speed are available.

Joints in Belting. Wherever possible, leather belts should be made endless with cemented laps. Endless belts cost less to maintain, wear longer and give better performance. Manufacturers of leather belting have belt-service departments convenient to all manufacturing centers. Textile belts and rubber belts are joined temporarily with fasteners when emergency warrants.

A—Preparing laps for endless single leather belts (Graton and Knight Co.).

B—Preparing laps for endless double leather belts.

C—Alternate method for laps of endless double leather belts.

D—Belt stretcher and clamp for cementing endless leather belts.

E—Wire hooks for round belting up to $3/8$ in. diameter.

F—Sewn joint for rubber belt.

G—Flanged and bolted joint.

H—Moxon's fastener (Great Britain).

J—Laminated leather belting.

K—Link leather belt.

L—Link leather belt for rounded pulleys.

M—Link leather V-belt.

N—Flat-link leather belt.

O—Texrope laminated-construction V-belt (Allis-Chalmers).

P—Texrope twin-cable-cord heavy-duty V-belt (Allis-Chalmers).

Q—Moran steel-belt coupling for round leather belts up to 1 in.

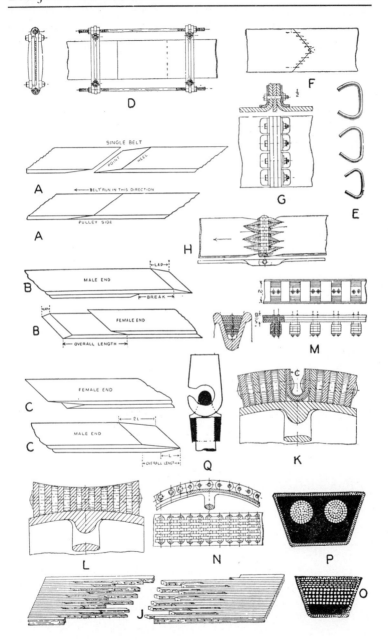

CLASS IV.   BASIC MECHANICAL MOVEMENTS

Section 34f.   Belt Joints and Fasteners

A—Stitched and cemented belt joint.

B—Double belting.

C—Laced lap joint.

D—Butt joint with apron piece.

E, F—Laced butt joints.

G—Pin and link joint.

H—Riveted and cemented joint.

J—Lacrelle's fastening.

K—Flattened-hook joint.

L—Blake's stud fastener.

M—Hook and eye joint.

N—Hinge fastening.

O—Metal spikes used as belt fasteners.

P—Sonnenthal's belt screw.

Q—Greene's belt stud; the ends of the belt are butted
together and holes are punched through simultaneously; the fasteners are put in as shown at a, twisted
90°; the band is next pulled straight and flattened
with a hammer; the completed joint is shown at b.

R, S—Pin and link joints.

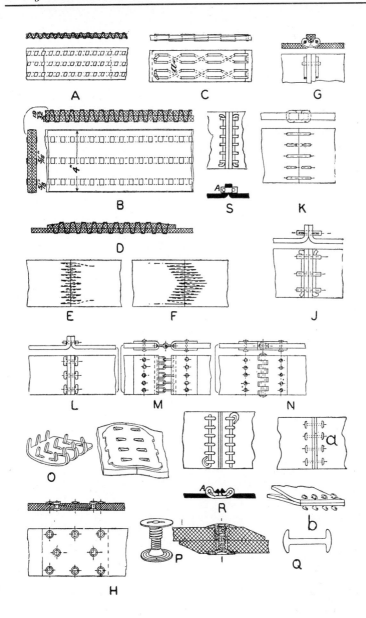

A

C

G

B

S

K

D

J

E

F

L

M

N

O

a

b

H

P

R

Q

CLASS IV.   BASIC MECHANICAL MOVEMENTS

Section 34g.   Texrope and Leather-Rope Drives

A—Variable-pitch sheave; two grooves; low speed (Allis-Chalmers).

B—Variable-pitch sheave; two grooves; high speed.

C—Variable-pitch sheave; single groove; low speed.

D—Variable-pitch sheave; single groove; high speed.

E—Texrope Vari-Pitch speed changer at low speed.

F—Texrope Vari-Pitch speed changer at high speed.

G—Square leather-rope drive (Tullis).

H—Core leather-rope with covered washer.

J—V-core leather rope.

K—V-shaped leather rope.

L—Round leather rope.

M—Twisted leather rope (up to $3/4$ inch diameter).

N—Sheave for twisted leather rope.

O—Square-core leather rope.

P—Double control sheaves arranged together to drive with twice-single variable-pitch range (Allis-Chalmers).

Q—Round leather flexible sectional belt.

R—Veelos laminated adjustable V-belt (Manheim Manufacturing and Belting Co.).

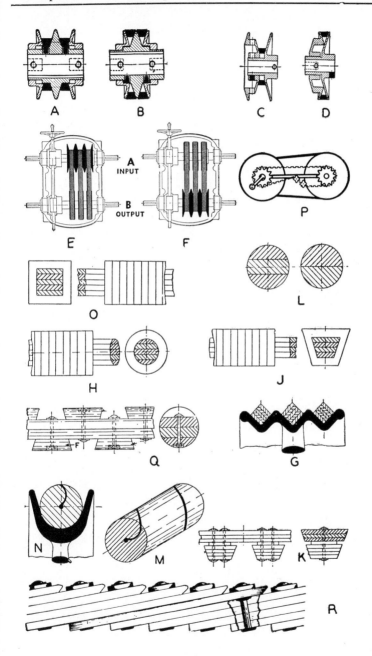

A

B

C

D

A INPUT

B OUTPUT

E

F

P

O

L

H

J

Q

G

N

M

K

R

CLASS IV.   BASIC MECHANICAL MOVEMENTS

Section 34h.   Textile-Rope Drives

A—Form and proportion of pulley grooves; the unit diameter equals the diameter of the rope.

B—Rope drive with separate textile ropes (English system).

C—Rope drive with continuous rope (old American system).

D—Pulley for hand rope.

E—Guide pulley; the rope rests on the bottom of the groove.

F—Canted groove for oblique drive.

G—Canted groove, showing advancing sections of the rope; permissible with rope drive.

H—Groove for double obliquity.

J—Short splice for textile ropes.

K—Long splice for textile ropes.

L—Kortum's rope fastening for coupling textile ropes.

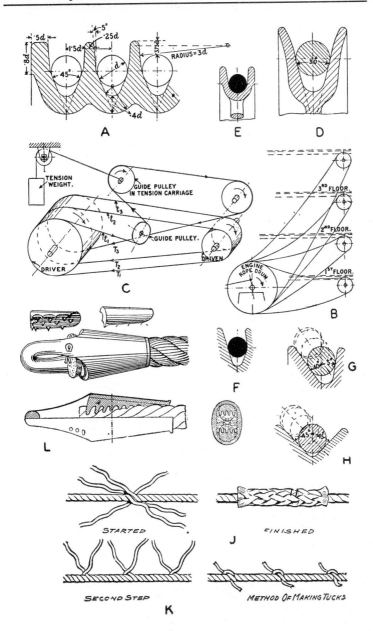

CLASS IV.   BASIC MECHANICAL MOVEMENTS

Section 34j.   Textile-Rope Drives

A—Improved American continuous rope drive with winder (Medart).

B—Split driven pulley; section of figure C on line A, B.

C—Split driven pulley.

D—Split driven pulley; section of figure C on line MN

E—Flywheel rim with grooves for a rope and teeth for barring a pawl.

F—Section of the rim of an engine-rope drum.

G—Mill driven by cotton ropes (Great Britain).

H, J—Pulley acting as flywheel of an engine; the wheel is 32 feet in diameter and has thirty-six grooves for ropes 1⅜ inches in diameter.

K—Travelling crane driven by a cotton rope (Great Britain).

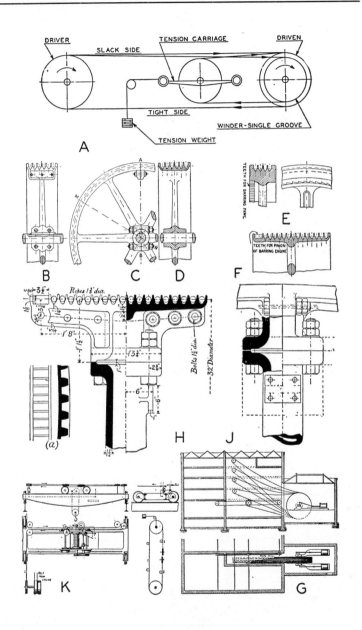

CLASS IV.   BASIC MECHANICAL MOVEMENTS

Section 34k.   Pitch-Chain Drives

A, B—Simplest forms of single flat-link pitch chain without rollers; used on speeds up to 800 feet per minute; shown on sprocket.

C—Double flat-link pitch chain.

D—Benoit roller chain.

E—Baldwin's chain; easy to remove a link.

F, G—Renold's chain shown on sprocket (Great Britain).

H—Morse silent chain good for high speeds.

J—Frictionless rocker joint for H.

K—Morse roller chain in double width; it is made also in up to quadruple widths.

L—Morse roller chain with extended pin; used on timing devices and in conveying.

M—Brampton block chain.

N—Triple roller chain (Boston Gear Works).

O—Ladder chain for light drives.

P—Appleby's adjustable chain for bicycles and light drives.

Q—Appleby's machine chain.

R—Hornsby's chain link.

S—Hall's detachable pitch chain.

T—Ewart's detachable link belt.

U—Drip-feed lubrication of chain drives.

V—Oil-bath lubrication.

W—Sprocket-wheel and pitch-chain assembly for very light duty.

X—Link belt and sprocket.

Y—Toothed linked chain.

Z—Link chain and sprocket (German).

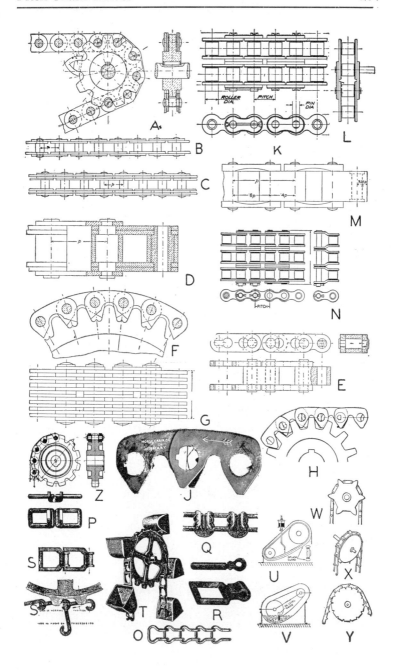

CLASS IV. BASIC MECHANICAL MOVEMENTS

Section 35a. Commercial Wire Ropes

In designating wire-rope construction, it is customary to state first, the number of strands; second, the number of wires in a strand; third, the kind of center or core whether fiber, hemp, wire strand or wire rope. When wire rope remains in a fixed position (such as in cables for suspension bridges) or where little bending is required, a wire core is desirable. For transmission of motion, flexibility over grooved pulleys is desirable and is secured by thinner wires and hemp or fiber cores.

A–3 × 7; fiber center.

B–6 × 7; fiber center.

C–7 × 7; wire center.

D–6 × 8; hemp center.

E–6 × 13; hemp center; filler wire.

F–6 × 16; fiber center; filler wire.

G–7 × 19; wire-strand center.

H–6 × 19; fiber center; two stranding operations.

J–6 × 19; hemp center; Seale patent.

K–6 × 37; fiber center; filler wire.

L–6 × 41; wire-rope center.

M–18 × 7; nonpinning type hoisting rope.

N–6 × 19; flexible; Seale patent; wire-rope center.

O–6 × 19; hemp center; Warrington patent.

P–6 × 19; hemp center; filler wire.

Q–8 × 19; hemp center; Seale patent.

R–8 × 19; fiber center; filler wire.

S–8 × 19; hemp center; Warrington patent.

T–6 × 19; wire-rope center; filler wire.

U–6 × 22; wire-rope center; filler wire.

V–6 × 31; fiber center.

W–8 × 19; fiber center; two stranding operations.

X–6 × 12; one hemp center.

Y–6 × 12; seven hemp centers.

Z–6 × 37; wire-rope center; Seale patent.

AA–6 × 37; fiber center; two stranding operations.

BB–6 × 37; hemp center; three stranding operations.

CC–6 × 24; seven hemp centers.

DD–6 × 42; seven hemp centers; most flexible; called "tiller" or "hand rope."

EE–3 × 37; wire center.

FF–Typical wire-rope center.

GG–Typical hemp or fiber center.

HH–Typical strand center.

JJ–Steel wires twisted into a single strand of nineteen wires.

KK–Steel wires twisted into a single strand of fifty-one wires.

LL–Armored wire rope; 6 × 19; fiber center; sometimes wire center; used under severe hoisting conditions, such as dredging and heavy steam-shovel work.

MM–Marline-covered rope; 5 × 19; hemp center; used for ship's rigging and hoisting service where moisture is encountered (American Chain and Cable Co.).

NN–Stone sawing strand; three wires twisted together.

OO, PP–Regular-lay (right and left) wire rope; wires in strands twisted together in one direction and strands twisted in opposite directions.

QQ, RR–Lang-lay (right and left) wire rope; wires in strands and strands twisted in the same direction

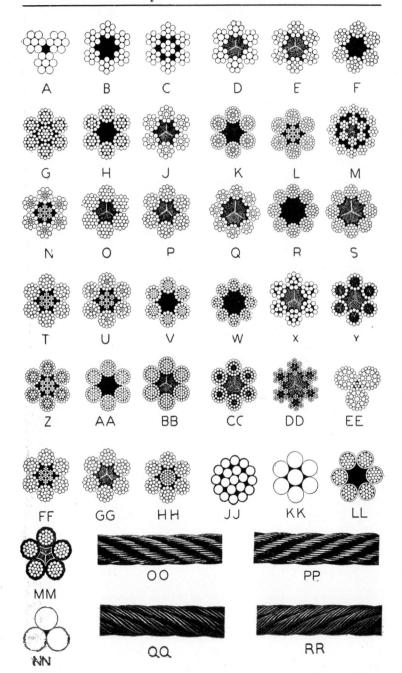

A B C D E F

G H J K L M

N O P Q R S

T U V W X Y

Z AA BB CC DD EE

FF GG HH JJ KK LL

MM OO PP

NN QQ RR

CLASS IV. BASIC MECHANICAL MOVEMENTS

Section 35b.   Wire-Rope Pulleys

A—Guide pulley or support; the bottom of the groove fits the wire rope.

B—Driving pulley; the bottom of the groove is lined with a material softer than metal, generally leather, but wood, rubber, etc., are also used.

C—Two-grooved lined pulley.

D—Clip pulley; the rim is divided into a series of toggles with pin joints at B and C causing the wire rope to be gripped with a force which varies with the tension of the rope (Fowler).

E—Differential wire-rope pulley; each groove is formed in a ring complete in itself; they are placed side by side with a combined clearance of ½ inch for six grooves; the rings adjust themselves automatically to all conditions of the load; the rope never slips in the groove, and the friction between rings and rim is sufficient for all purposes; a grease lubricator is shown at L: if ring slippage occurs, the flanges at F are tightened up; an elastic cushion C is placed between the flange and the pulley (Walker).

F—Mining car connected to wire cable by a screw grip G, or by knotting a chain around the cable to engage a projecting bar C.

G—Tramway-cable grip.

H—Shackle for wire rope.

J—Single-rope lined sheave.

K—Bolted steel line-grooved hoisting sheave (Medart).

L—Correct arc for the bottom of the groove to give intimate contact.

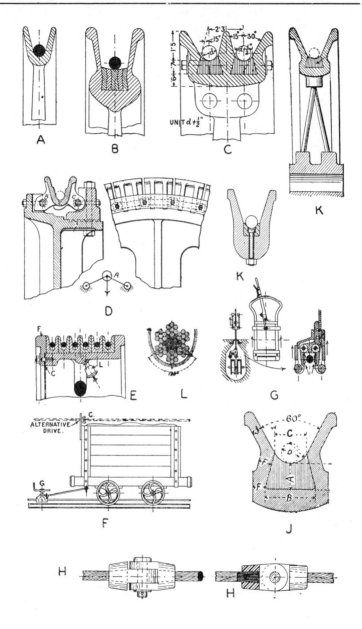

CLASS IV.   BASIC MECHANICAL MOVEMENTS

Section 35c.   Long-Distance Wire-Rope Transmission

A—Transmission without supporting pulleys.

B—Transmission with supporting pulleys; driving side is at the bottom.

C—Transmission with supporting pulleys; driving side is at the top.

D—Transmission over hilly ground.

E—Transmission with long-space intermediate pulleys.

F—Guide pulley.

G, H—Use of toothed gearing.

J—Mounted intermediate pulley.

K—Mounted pulley.

L—Angle-bar guide for mounting rope.

M—Driving friction drum for low speeds.

N—Tension producer for low speeds.

O—Rope-supporting underground haulage.

P—Grooved driving and counter pulleys.

Q—Patent rope clips and clamps (Bullivant).

R—Shackles for wire rope.

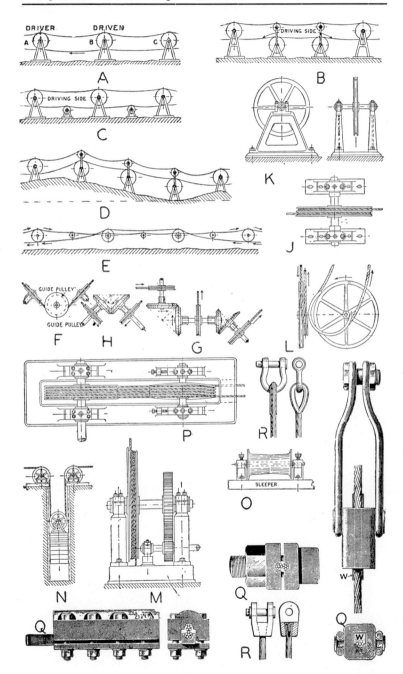

CLASS IV.   BASIC MECHANICAL MOVEMENTS

Section 36a.   Rudimentary Clutches

A–Ring friction clutch (Hill Clutch Co.).

B–Cone clutch; screw gear is used to operate this, as it is liable to "seize" or "grab"; there is considerable end pressure in the shell to be allowed for.

C–Friction-clutch face with V-grooves.

D–Cam clutch; self-acting clutch arms act on the pulley in one direction only; when the shaft motion is reversed, the pulley is free.

E–Clutch and gear; the clutch slides on the feathered shaft; the gear is thrown into motion by the bell-crank lever and runner.

F–Coil-grip friction clutch; the coil is a steel spring.

G–Internal-grip friction clutch; the internal ring is split at one side and expanded by the oval pin attached to the arm.

H–Compound-disc friction clutch with V-grip; each alternate disc is locked to the shaft and the intermediate discs to the drum (Hele-Shaw).

J, K, L, M, N–Roller and gravity-pawl clutches; they run free in one direction and grip instantaneously in the other direction.

O–Internal-cone friction clutch with leather or other lining.

P–Pin clutch.

Q–Friction pin clutch; a is a friction band that slips to prevent shock when the pins b and d are thrown in contact.

R–Cone friction clutch; preferred for connecting shafts with frequent and variable loads.

S–Friction-clutch bevel gear.

T–Spring friction clutch.

U–Double toggle-joint friction clutch.

V–Adjustable friction clutch with double-grip bearings.

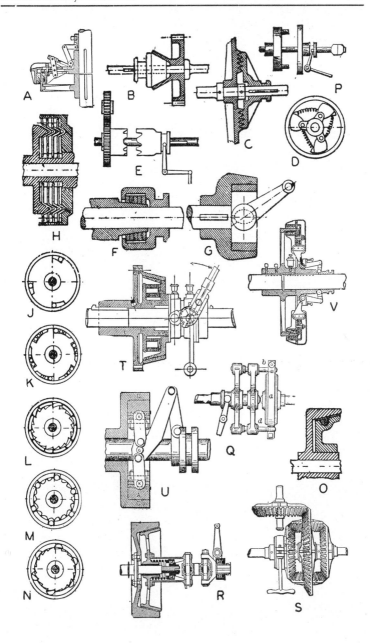

CLASS IV.   BASIC MECHANICAL MOVEMENTS

Section 36b.   Friction Clutches

A—Friction clutch; filled with oil (Hele-Shaw).

B—Friction clutch (Dohmen-Leblanc).

C—Friction clutch (Bagshaw).

D—Coil clutch; to be filled with circulating oil.

E—Disengaging clutch.

F—Slipping friction clutch.

G—Old-style automobile clutch with leather-faced cone.

H—Improved form of G; metal to metal; a predecessor of the disc clutch.

J—Friction clutch (Mather and Platt).

K—Shifting gear for clutches; the lever may be worked by hand, but a screw and hand wheel are sometimes used.

L—Lever for the shifting gear K; it is filled with brass blocks to increase the wearing surface of the forked end.

M—A strap which increases the wearing surface of the forked end of the lever for the shifting gear K.

N—Friction clutch (German design).

O—Old-style friction clutch.

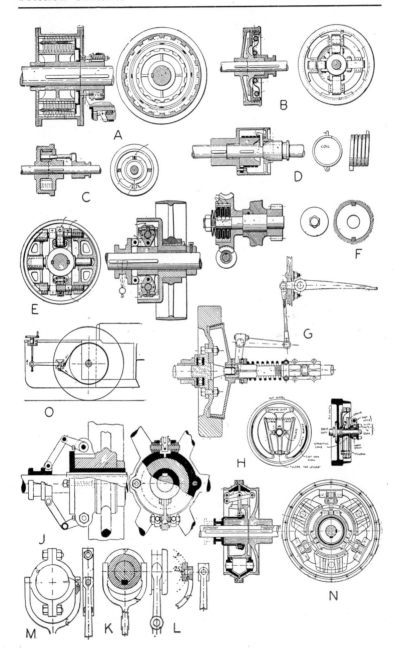

CLASS IV.   BASIC MECHANICAL MOVEMENTS

Section 36c.   Commercial Friction Clutches and Couplings

A—Twin-disc friction-clutch coupling; enclosed type with a spider mounted on the hub and a ball-bearing collar (Link-Belt Co.).

B—Twin-disc friction clutch with bronze collar.

C—Twin-cone friction clutch.

D—Multiple-disc friction clutch.

E—Steel safety collar.

F—Malleable safety collar.

G—Rigid ribbed compression coupling.

H—Keyless compression coupling.

J—Flexible coupling constructed of two cast-iron flanges and an intermediate flexible engaging member.

K—Flexible coupling, consisting of two cut-teeth sprocket wheels which are connected by a roller chain (Link-Belt Co.).

L—Mercury midget clutch, consisting of a driving member carried on the motor shaft, and a driven member, which drives the load by V-belt, or direct coupling; the driving member consists of two friction plates A, between which is a rubber gland B containing a small quantity of mercury C which, when at rest, occupies the ring-shaped cavity near the shaft; when revolved, the mercury is urged outwardly by centrifugal force, thus spreading the rubber gland axially, forcing the friction plates against the inner faces of the housing D, causing the driven member to revolve; available up to $1/4$ horsepower capacity.

M—Mercury heavy-duty clutch for motors up to 15 horsepower and for internal combustion engines (Mercury Clutch Co.).

N—Friction-clutch coupling (Link-Belt Co.).

O—Shifting lever for the clutch of figure N.

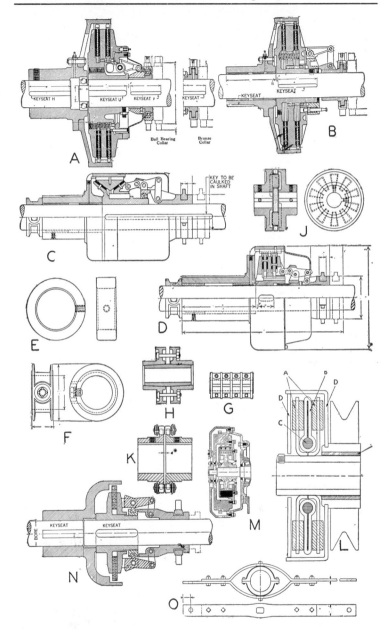

CLASS IV.   BASIC MECHANICAL MOVEMENTS

Section 36d.   Rigid-Type Shaft Couplings

A—Fairbairn's lap-box shaft coupling.

B—Half-lap shaft coupling.

C—Cased-butt shaft coupling.

D—Plain butt-muff shaft coupling.

E—Friction-clip (muff-type) shaft coupling.

F—Split-box shaft coupling.

G—Flanged shaft coupling.

H—Seller's cone shaft coupling.

J—Bulter's frictional muff coupling.

K—Marine tail shaft coupling.

L—Combination box and flange (Archibald Sharp).

M, N—Combination pulleys and couplings (German).

O—Flange coupling with filler piece (German).

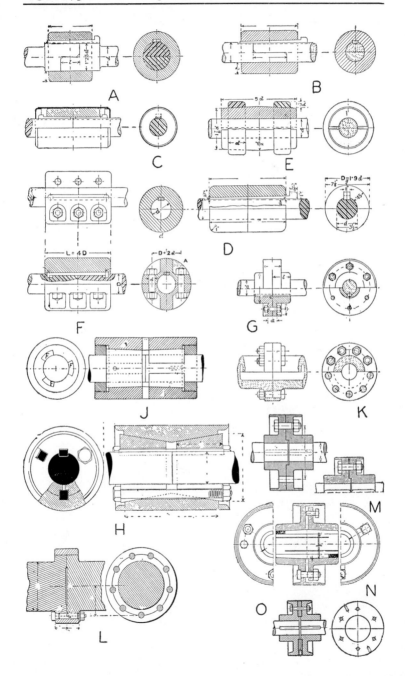

CLASS IV.   BASIC MECHANICAL MOVEMENTS

Section 36e.　Flexible Shaft Couplings

A—Brotherhood's flexible coupling; the hollow casting D is bolted to the shaft and A; the corrugated disc E is bolted to the shaft and B; the shafts bolt together on a spherical surface C which permits carrying an end thrust as in marine-propeller shafts; this design cannot well support axial tension; the coupling is comparatively large in diameter.

B—Alley's flexible coupling with solid flanges; the left flange has a spherical projection K which fits into two half-discs J fastened to the right flange and the pivot L bears against the shaft G; there are steel bushes in the left flange to receive the barrel-shaped ends of the bolts, which are a loose fit radially; tension or compression in an axial direction is fully resisted and the coupling is comparatively small in diameter.

C—Hopkinson's flexible coupling for light torques.

D—Oldham's coupling for two shafts slightly out of line; if the shafts are considerably out of line, a universal joint is used.

E—Leather-plate coupling; very flexible (German).

F—Claw coupling for large and slow-moving shafts; one-half of the coupling is shown with a feather or sliding key, to permit easy disconnection.

G—Claw coupling; the two halves are easily put together, but rotation must take place always in the same direction.

H—Transmitter; friction-type coupling (Hele-Shaw).

J—Flexible link coupling; the links are made of leather.

K, L—Leather bolt flexible coupling (German).

M—Flexible coupling with rubber sleeves (German).

N—Flexible coupling, (Wülfel Iron Works of Hanover, Germany).

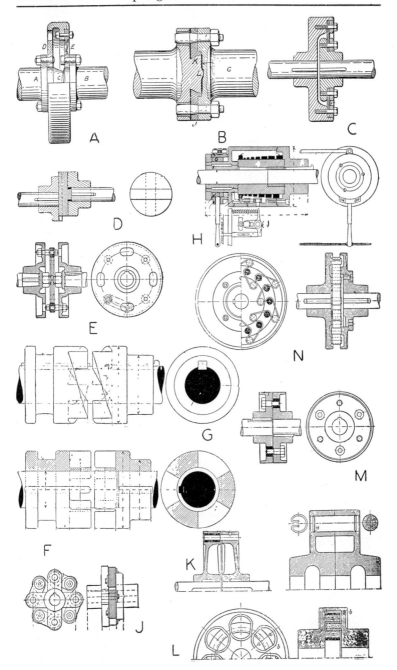

CLASS IV.   BASIC MECHANICAL MOVEMENTS

Section 36f.   Shaft Couplings and Friction Clutches*

A—Band coupling (J. M. Voith).

B—Band coupling with two bands (J. M. Voith).

C—Band type coupling; very elastic (Peniger Machine Co.).

D—Hildebrandt coupling (rigid type).

E—Zodel-Voith band coupling.

F—Friction clutch coupling (G. Polysius).

G—Centrifugal coupling.

H—Hill friction-clutch coupling (Wülfel Iron Works).

J—The X-coupling; friction-clutch type.

K—Friction-clutch couplings (Louis Schwarz & Co.).

L—Uhlhorn friction-clutch coupling.

M—Rutsch friction-clutch coupling (Lohmenn & Stolter-fohet).

N—Friction-clutch coupling (Louis Schwarz & Co.).

---

*These flexible couplings and friction clutches were in use in Germany before World War II.

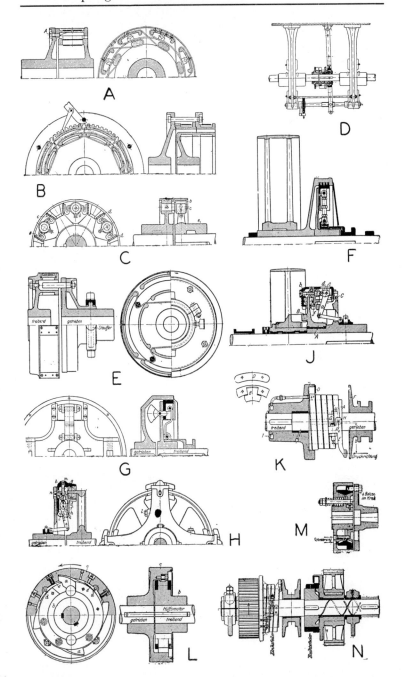

CLASS IV.　BASIC MECHANICAL MOVEMENTS

Section 36g.　Commercial Flexible Couplings

A—Flexible disc coupling (Thomas).

B—Grundy flexible coupling; leather-disc lugs fit into openings cast into the two flanges.

C—Chain jaw-clutch coupling (Morse).

D—Double slider coupling (Jones Foundry and Machine Co.).

E—Detail of flexible pin coupling (Smith and Serrell, Inc.).

F—Fast's gear coupling. (Bartlett Hayward Co.).

G—Universal giant friction-clutch cut-off coupling (Wood's Sons Co.).

H—Francke laminated pin flexible coupling (Foote Bros.).

J—Rubber-cushion coupling (Foote Bros.).

K—Leather disc between the flanges, which provides flexibility and insulation (Wood's Sons Co.).

L—Steel rubber-bushed pins which absorb shock and give the coupling flexibility and insulation (Wood's Sons Co.).

M—Roller-chain coupling with plastic casing; it is considered good practice to enclose and lubricate flexible couplings when speeds of 500 revolutions per minute or greater are required (Link-Belt Co.).

N—Bend-type coupling.

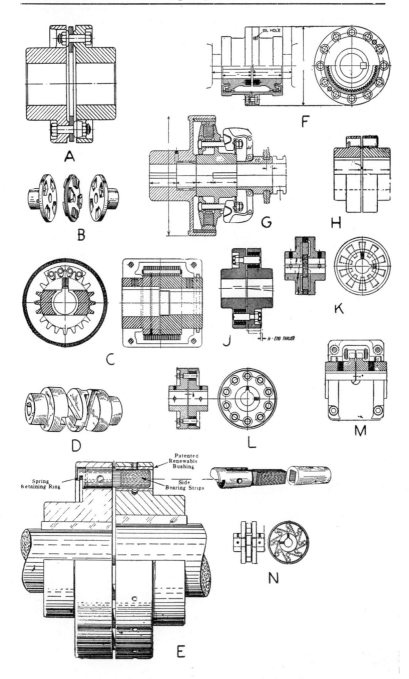

A

B

F

G

H

C

J

K

D

L

M

E

N

Patented
Renewable
Bushing

Spring
Retaining Ring

Side
Bearing Strips

CLASS IV.   BASIC MECHANICAL MOVEMENTS

Section 36h.   Commercial Universal Joints

A, B—Hooke's joints; first commercial design (Great
   Britain).

   C—Universal joint (Boston Gear Works).

   D—Universal joint; maximum working angle 36° (Brooks
   Equipment Co.).

   E—Universal joint; approximate operating angle 17°
   (Brooks Equipment Co.).

   F—Universal joint; operating angle 30° (Brooks Equipment Co.).

   G—Bronze universal joint; operating angle 36° (Brooks
   Equipment Co.).

   H—Bolt-type universal joint; operating angle 36° (Brooks
   Equipment Co.).

   J—Slip universal joint for use where one shaft has unusual longitudinal movement, e.g., for rising stem
   valves (Brooks Equipment Co.).

   K—Shaft slip coupling for unusual axial play (Brooks
   Equipment Co.).

   L—Semiuniversal coupling similar to Oldham's (Foote
   Bros.).

   M—Weiss universal joint; a number of steel balls inserted
   in intersecting races, cut in the two joint members
   and transmit motion, permitting axial movement of
   the shafts.

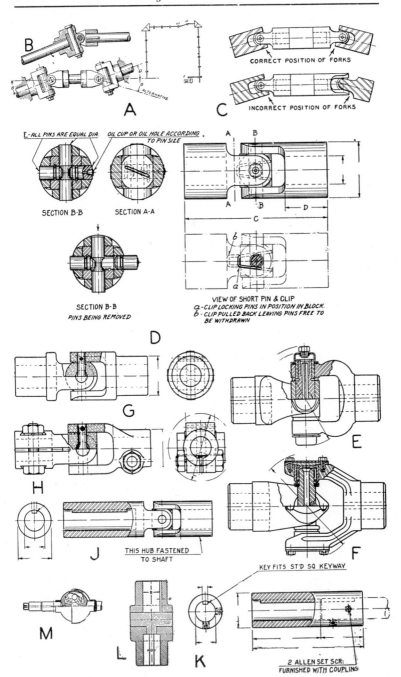

B

A

C CORRECT POSITION OF FORKS

INCORRECT POSITION OF FORKS

E-ALL PINS ARE EQUAL DIA.

OIL CUP OR OIL HOLE ACCORDING TO PIN SIZE

SECTION B-B   SECTION A-A

SECTION B-B
PINS BEING REMOVED

VIEW OF SHORT PIN & CLIP
a-CLIP LOCKING PINS IN POSITION IN BLOCK.
b- CLIP PULLED BACK LEAVING PINS FREE TO
BE WITHDRAWN

D

G

H

E

F

J   THIS HUB FASTENED
TO SHAFT

KEY FITS ST'D SQ KEYWAY

M   L   K

2 ALLEN SET SCR:
FURNISHED WITH COUPLING.

CLASS IV.   BASIC MECHANICAL MOVEMENTS
Section 36j.   Rudimentary Angular Couplings

A—Flexible angular coupling for light work.

B—Angular coupling for shafts to replace bevel gearing.

C—Angular coupling for shafts at any angle (shown at 90°), consisting of four crank pins sliding and revolving in holes bored in the ends of shafts (Hobson's patent).

D—Universal coupling.

E—Elastic coupling for shafts at very slight angle.

F—Pin and slot shaft coupling.

G—Elastic coupling.

H—Angular coupling on Hooke's principle.

J—Sliding-contact shaft coupling with a bar sliding in two yokes on the offset shaft.

K—Angular crank-pin motion converted to rectilinear motion; rotating shaft A carries crank pin E; an arm with sleeve D is jointed to the yoke and sliding rod B.

L—Angle-shaft coupling.

M—Universal joint with one cross link; it operates at angles less than 45°.

N—Double-link universal joint for greater angles.

O—"Almond" angle-shaft coupling; yoke links G are pivoted to sockets at the ends of shafts A, B and to the 90° arms E on the sleeve which slides freely on the fixed shaft D; the sockets at F are ball joints; the angle of the shaft may vary within limits.

P—Sliding-sleeve angle-shaft coupling.

Q—Universal angle-shaft coupling on Hooke's principle; each shell carries a double-pivoted trunnion ring, the connecting link being pivoted at each end to the rings.

R—Ball and socket universal joint.

S—Goubet's universal shaft coupling; trunnion ring C is recessed in ball D; each shell is alike and individually a universal joint for 45°; together, they will operate shafts A at 90°.

T—Universal joint with gimbal ring.

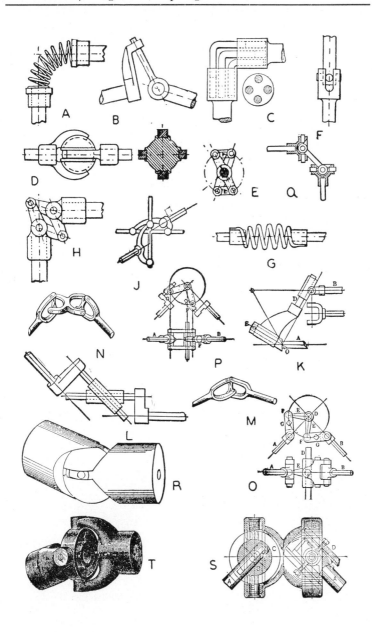

CLASS IV.   BASIC MECHANICAL MOVEMENTS

Section 36k.   Commercial Angular Shaft Transmission

A—Flexible gear turret; it eliminates special angle bevel gears; one bracket is adapted to swing in an arc of 190° to suit the varying angle; use of two brackets allows both shafts to remain in the selected angle (Brooks Equipment Co.).

B—Brooks flexible reversing-gear turret.

C—Brooks flexible gear turret bracket; pedestal type.

D—Brooks flexible gear turret bracket; strap type.

E—Brooks straight-line reversing-gear box for reversing the shaft direction when the straight center line of shafts is to be maintained.

F—Consolidated hinged joint for use in 90° position but allowing 2° deviation in either direction; adapted to connect an auxiliary or branch shaft to a main-line shaft if a shaft extension is installed; it may also be hooked up to reverse the direction of the shaft; the shaft diameter is $\frac{1}{2}$ to $2\frac{1}{2}$ inches (Brooks Equipment Co.).

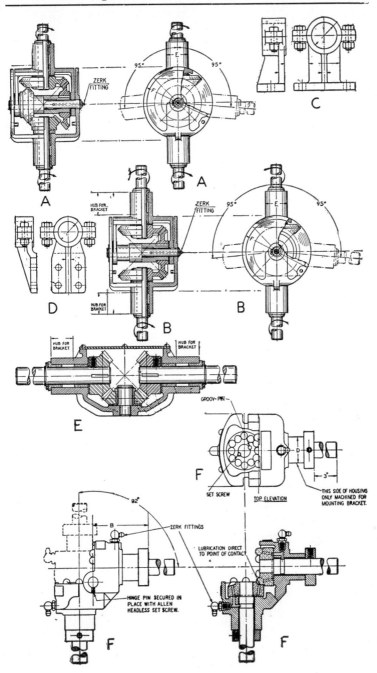

CLASS IV.   BASIC MECHANICAL MOVEMENTS

Section 361.   Rudimentary Brakes

A—Lever and block brake with wooden, cast-iron, or similar brake lining.

B—Lever and strap brake; the strap is usually faced with wood, leather or other brake-lining material.

C—Compound lever and block brake.

D—Lever and double-block brake.

E—Disc brake; arrangements must be made for axial pressure.

F—Fan brake for small devices.

G—Spring brake for light duty.

H—Rope brake or grip with toggle motion.

J—Rope brake.

K—Rope brake with car-lever grip.

L—Block brake with eccentric lever action.

M—Strap and screw brake; faced with brake lining.

N, O, P—Car brakes.

Q—Combined lever and strap brake.

R—Shaft grip or brake.

S—Centrifugal brake or clutch.

T—Compound brake with three segments.

U—Compound-bar brake.

V—Compound-ring brake.

W—Wedge and split-ring brake.

X—Hollow drums, with radial pockets, partially filled with water, mercury, etc., which retards the motion.

Y, Z—Railway-car brakes.

AA—Hydraulic buffer.

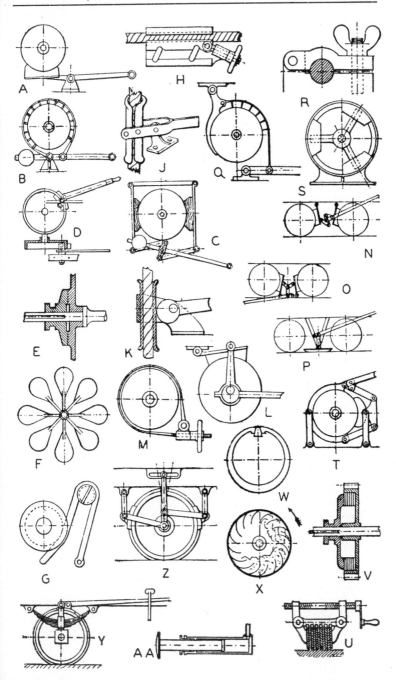

CLASS IV.   BASIC MECHANICAL MOVEMENTS

Section 36m.   Contracting and Expanding Mechanisms

A, B, C—Parallel-bar expanding grilles or gates.

D—Venetian blind; used for sliding doors.

E—Perforated-bar and hooked-rod hanger.

F—Expanding link device with four guides.

G—Lazy tongs.

H—Expanding basket.

J—Expanding socket.

K—Expanding grating for a gate.

L—Expanding legs for a tripod.

M—Expanding screen.

N—Expanding pipe stopper with rubber ring

O, P, Q—Expanding pulleys or wheels.

R—Expanding pipe grip.

S—Expanding gate.

T—Expanding tripod.

U—Car bumper.

V—Expanding socket formed of spring wire.

W—Expanding chuck or mandrel.

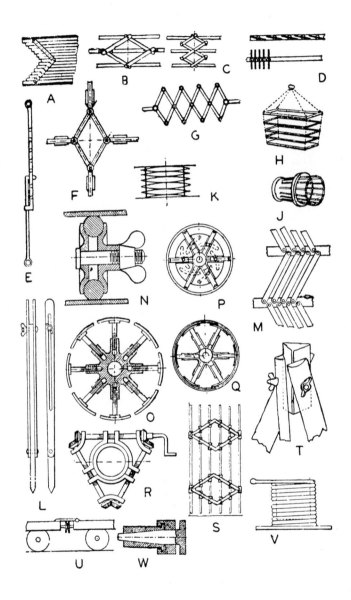

CLASS IV.   BASIC MECHANICAL MOVEMENTS

Section 36n.   Contracting and Expanding Mechanisms

A—Telescopic-ram hydraulic lift.

B—Thorburn's tube expander.

C—Gasometer.

D—Timm's expanding boring tool; it has a central cone and three or more diagonal feathers, sliding in dovetail grooves in the central cone.

E—Expanding mandrel with three parallel feathers expanded by a central bolt having two equal cones.

F—Expanding reamer; split up into three parts as far as the end of the bolt.

G—Addy's expanding collar consisting of two rings with spiral adjacent faces, so that by revolving them, they separate to the amount of the pitch; the collar expands longitudinally, but not diametrically.

H—Expanding split reamer or mandrel with taper screw.

J—Expanding sleeves or collars screwed one on the other.

K—Expanding collet split in three parts.

L—Expanding lever.

M—Rich's expanding mandrel with a square helix parallel outside, but tapered in the bore and sliding on a tapered mandrel.

N—Expanding mandrel; the parallel bushing is split alternately from each end tapered in the bore to fit the tapered mandrel.

O—"The Mannesmann Process" of seamless tube fabrication; A, corrugated rolls; B, guide tube; B'' hot bar; it is expanded by mandrel.

P—Seamless tube fabrication; rolling a solid bar between a pair of angular-axled disc rollers opens a cavity within the bar which is further expanded by a second pair of disc rollers; the rolling of the tube between the discs pushes the tubular bar over a revolving conical mandrel.

Q—Taper tube rolling.

R—Multiple butterfly valve ventilator.

S—Organ bellows.

T—Spiral-type bimetallic thermostat; the contact is made and broken by a mercury switch.

U—Steam trap using charged metal bellows for power element (Fulton Sylphon Co.).

V—Bimetallic thermostat; curved-strip type.

W—Bimetallic thermostat; straight-strip type.

X—Diaphragm-type thermostat; volatile liquid expands and contracts with change of temperature.

Y—Direct-expansion type thermostat.

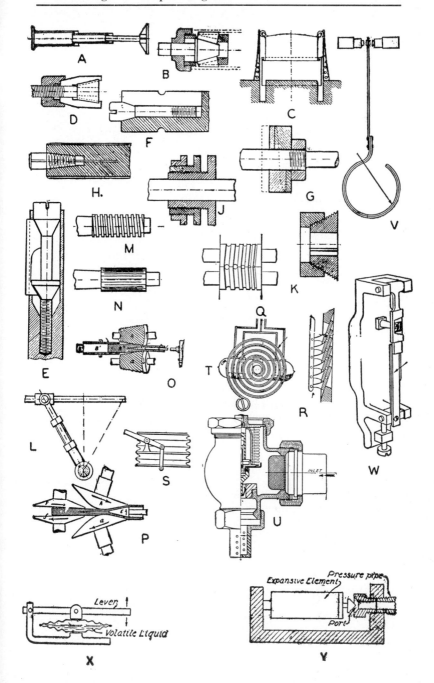

CLASS IV.   BASIC MECHANICAL MOVEMENTS

Section 36o.   Springs

A—Open helical compression spring; round wire.

B—Open helical compression spring; square wire.

C, D—Volute springs; conical.

E—Double-volute chair spring.

F—Open helical tension spring.

G—Close helical spring.

H—Spindle shape open or closed helical tension spring.

J—Open or closed parallel-helical spring with coned ends.

K—Fixed flat spring.

L, M—Sear springs.

N—Wire spring.

O—Flat spiral spring.

P—Plate spring.

Q—Rubber tension spring.

R—Ribbon torsion spring.

S—Compound rubber-disc spring.

T—Air cushion or spring piston; dash pot.

U—Laminated-plate wagon spring.

V—Loop spring.

W—Compound dished-disc or bent-plate spring.

X—Split-ring spring.

Y—Flat spiral spring used as clock spring or coil spring.

Z, AA, BB—Spring washers.

CC—Spindle-shaped compression spring.

DD—Flat spiral spring used on piston rings.

EE—Wood springs of ash.

FF—Carriage spring with link suspension.

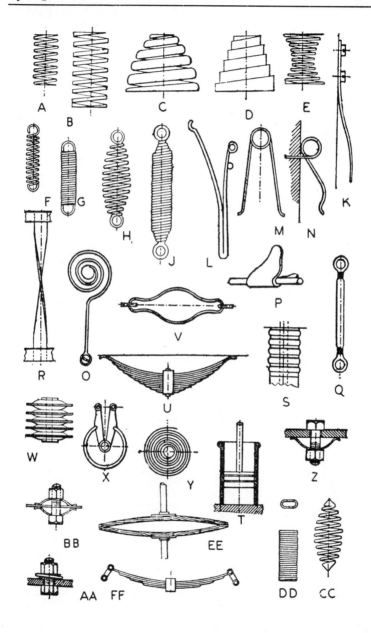

CLASS V.   ELEVATORS, DERRICKS, CRANES, CONVEYORS

Section 37a.   Gantry Cranes

A—Gantry crane; similar to an overhead traveling crane, except that the bridge for carrying the trolley or trolleys is rigidly supported on two or more movable legs running on fixed rails, or other runway.

B—Semigantry crane with one end of the bridge rigidly supported on one or more movable legs running on a fixed rail or runway, the other end of the bridge being supported by a truck running on an elevated rail or runway.

C—Cantilever gantry crane; the bridge girders or trusses are extended transversely beyond the crane runway on one or both sides; its runway may be on the ground or elevated.

D—Portal crane; without trolley motion; it has a boom attached to a revolving crane mounted on a gantry; the boom can be raised or lowered at its head (outer end); it may be fixed or mobile.

E—Tower portal crane for hoisting and swinging loads over high obstructions; mounted on a fixed or mobile tower-like gantry, e.g., by a revolving mast or turntable, with or without an opening between the legs of the gantry.

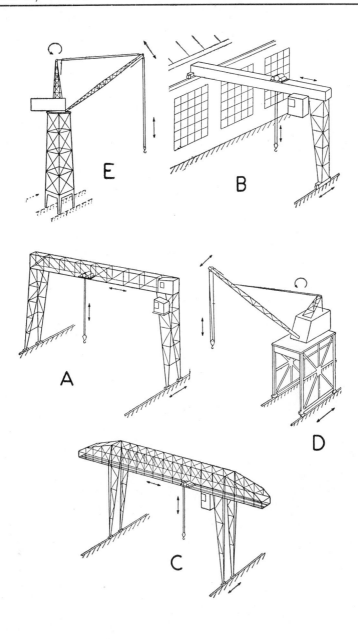

CLASS V.   ELEVATORS, DERRICKS, CRANES, CONVEYORS

Section 37b.   Cranes

A—Hammer-head crane; a rotating counterbalanced cantilever equipped with one or more trolleys and supported
by a pivot or turntable on a traveling or fixed tower.

B—Locomotive crane consisting of a self-propelled car, operating on a railroad track, on which is mounted a
rotating body supporting the power-operated mechanism, together with a boom which can be raised or lowered at its head (outer end), from which end a wire
rope is led for raising or lowering the load.

C—Crawler crane of the locomotive-crane type mounted on
a tractor frame instead of on a railroad car, using tractor or caterpillar belts or treads for locomotion in any
direction.

D—Overhead traveling crane on a pair of parallel elevated
runways, adapted to lift or lower a load and to carry it
horizontally and parallely or at right angles to the runways; it consists of one or more trolleys, operating on
the top or bottom of a bridge, which consists of one or
more girders or trusses mounted on trucks operating on
the elevated runway.

E—Wall crane having a jib with or without a trolley and
supported from a side wall or from line of columns of a
building so as to swing through a half circle only; it is
usually of the traveling type, in which case, it operates
on a runway attached to the side wall or columns.

F—Fixed jib crane with a vertical member supported at top
and bottom; a horizontal revolving arm carrying a trolley extends from the vertical member.

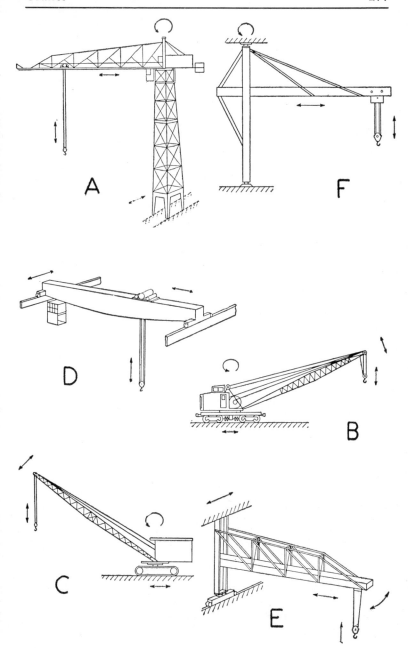

A

F

D

B

C

E

CLASS V.   ELEVATORS, DERRICKS, CRANES, CONVEYORS

Section 37c.   Cranes and Derricks

A—Fixed pillar crane with a vertical member held in position at the base to resist overturning moment and a constant-radius revolving boom supported at the outer end by a tension member.

B—Fixed pillar jib crane with a vertical member held at the base and a horizontal revolving arm carrying a trolley.

C—A-frame derrick in which the boom is hinged from a cross member between the bottom ends of two upright members spread apart at the lower ends and united at the top; the upper end of the boom is secured to the upper junction of the side members and the side members are braced or guyed from the junction point.

D—Breast derrick without a boom; the mast consists of two side members spread farther apart at the base than at the top, tied together at top and bottom by rigid members, the top held from tipping by guys, and the load raised and lowered by ropes through a sheave or block secured to the cross piece.

E—Shears with winch or tackle blocks.

F—Portable steam derrick on swiveled platform balanced by the boiler.

G—Swing-boom crane with traveling truck and trolley lift; the boom revolves on radial rollers.

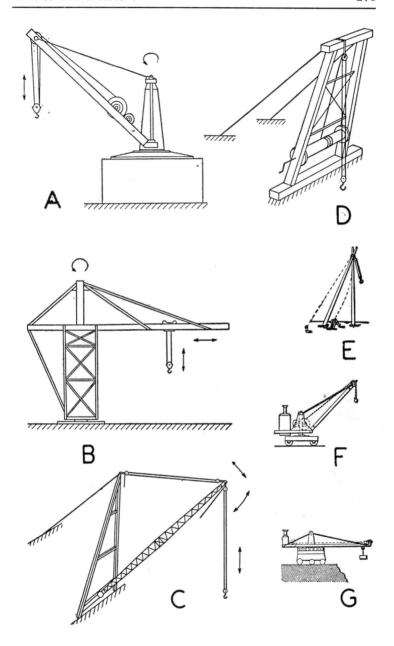

A

D

B

C

E

F

G

CLASS V.   ELEVATORS, DERRICKS, CRANES, CONVEYORS

Section 37d.   Cranes and Derricks

A—Gin-pole derrick; consists of a mast, with guys from its top so arranged as to permit leaning the mast in any direction, the load being raised or lowered by ropes leading through sheaves or blocks at the top of the mast.

B—Fixed guy derrick, the mast can be rotated; it is supported in vertical position by three or more guys and a boom whose bottom end is hinged or pivoted to move in a vertical plane, with lines between the head of the mast and the head of the boom for raising and lowering the boom, and lines from the head of the boom for raising and lowering the load.

C—Stiff-leg derrick similar to a guy derrick except that the mast is supported or held in place by two or more stiff members capable of resisting tensile or compressive forces; usually sills connect the lower ends of the two stiff legs to the foot of the mast.

D—Traveling derrick; double trolley installing a heavy gun on a battleship.

E—Elevator tower with inclined boom; a bucket is hoisted to the trolley by a double tackle, drawn up the incline and the load is automatically dumped into a car.

F—Mast and gaff hoist for unloading barges; either a portable boiler and engine hoist or electric motor supplies the power.

G—Horizontal boom tower with traversing trolley and automatic shovel bucket.

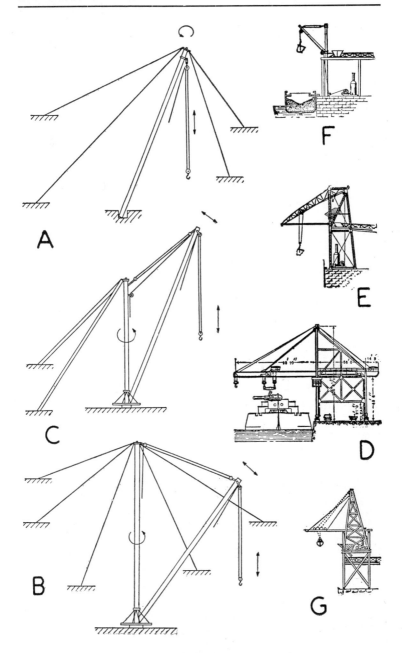

A

C

B

F

E

D

G

CLASS V.   ELEVATORS, DERRICKS, CRANES, CONVEYORS

Section 37e.   Commercial Cranes and Unloaders

A, B, C—Miscellaneous cantilever design of rotary cranes.

D—Boston-type unloader for barges.

E—Cantilever rotary crane mounted on a gantry for unloading freight cars.

F—Cantilever rotary crane mounted on a semigantry for unloading freight cars.

G—Gantry crane.

H—Cantilever gantry crane.

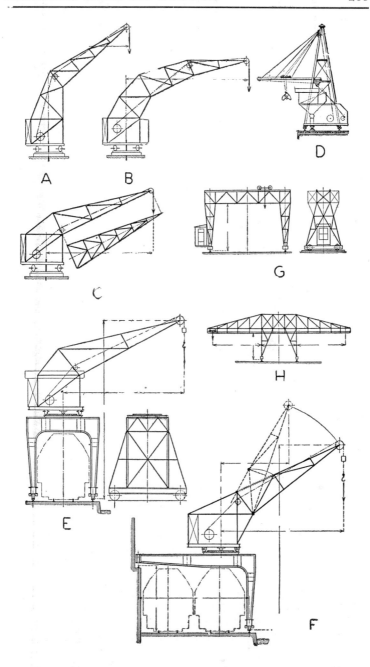

A   B   D

C

G

E   H

F

CLASS V.   ELEVATORS, DERRICKS, CRANES, CONVEYORS

Section 37f.   Commercial Wire-Rope Slings

A—Equalizing grommet cradle sling handling a casting (Roebling).

B—Equalizing grommet sling handling a locomotive.

C—Sling handling pig iron.

D—Six-part cradle sling (Roebling).

E—"Flatweave" thimble (Roebling).

F—Choker hook (Roebling).

G—Sling for lifting box cars at the end sill (Roebling).

H—Sling for lifting gondolas and flat cars (Roebling).

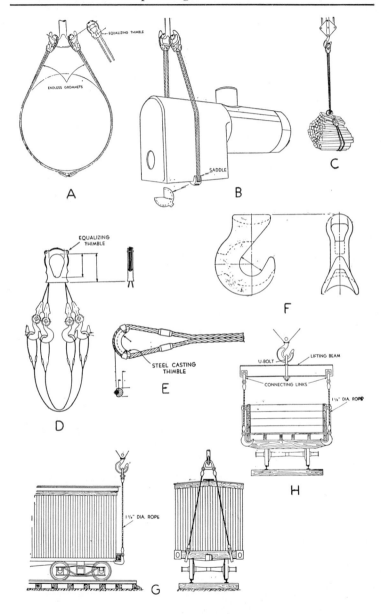

EQUALIZING THIMBLE

ENDLESS GROMMETS

A

SADDLE

B

C

EQUALIZING
THIMBLE

D

STEEL CASTING
THIMBLE

E

F

U-BOLT   LIFTING BEAM

CONNECTING LINKS

1⅛" DIA. ROPE

H

1¼" DIA. ROPE

G

CLASS V.   ELEVATORS, DERRICKS, CRANES, CONVEYORS

Section 37g.   Commercial Wire-Rope Slings

A—Sling handling a boiler (Roebling).

B—Sling for lifting passenger cars under cross-bearer (Roebling).

C—Sling handling the front end of a locomotive (Roebling).

D—Sling handling the rear end of a locomotive (Roebling).

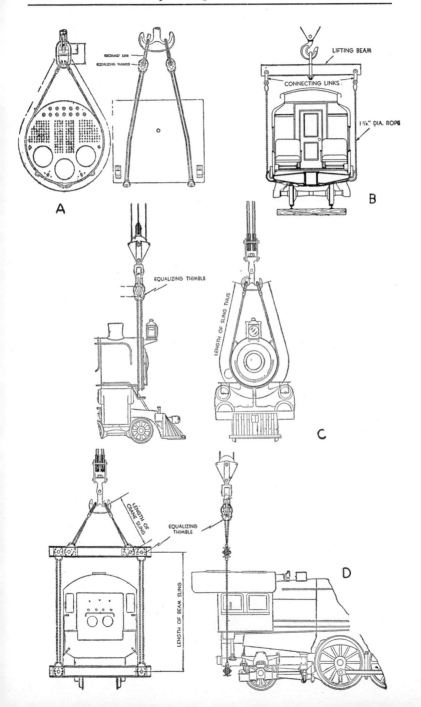

A

B

C

D

CLASS V.   ELEVATORS, DERRICKS, CRANES, CONVEYORS

Section 37h.   Fundamental Lifting Grabs and Clamps

*A—Plate clamp; see detail U.

B—Load beam; it will remain practically horizontal if unequally loaded.

C—Sling for marble slabs.

D—Lifting magnet for steel castings or parts.

E—Movable hook beam.

F—Lugged platform.

G—Round grip tongs.

H—Notched beam.

J—Bale and clamp hook.

K—Hand clamp box.

L—Wedge tongs.

M—Hairpin holder for tires on a conveyor system.

N—Hand clamp sheet.

O—Double and single tongs.

P—Barrel roll-over lifting grab; for examination or painting.

Q—Open-side lifting hooks for pipes, rods on a conveyor or hoist.

R—Barrel-end lifting hooks for hoist or conveyor system.

S—Motor-driven roll-over; similar to P.

T—Beam and sling for hoist or conveyor systems.

U—Merrill positive plate-lifting clamps.

V—Merrill positive drum and container clamp.

W—Merrill positive clamp handling angles and structure assemblies; 5 to 1 safety factor.

X—Method of lifting a pipe vertically.

Y—Method of lifting a pipe horizontally.

Z—Hoisting a flat plate with a rope sling.

AA—Hoisting a small plate on edge.

*Sketches A through S have been reprinted by permission of "Mill and Factory" magazine.

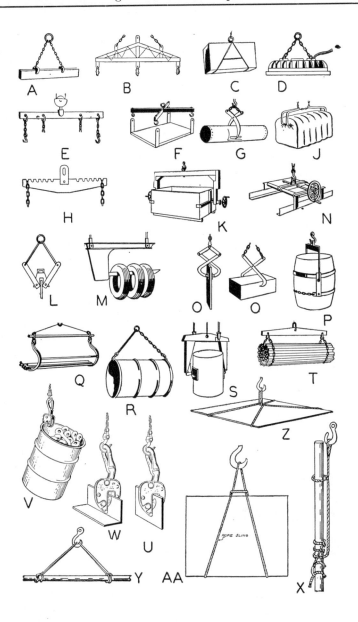

A
B
C
D
E
F
G
J
H
K
N
L
M
O
O
P
Q
R
S
T
V
W
U
Z
Y
AA
X

CLASS V.   ELEVATORS, DERRICKS, CRANES, CONVEYORS

Section 37j.   Rope Knots and End Fittings

A—Simple overhand knot.

B—Double-twist knot.

C—English knot.

D—Round turn and hitch.

E—Moorish knot.

F—Lark's head knot.

G—Simple boat knot.

H—Timber hitch.

J—Builder's knot.

K—Double Flemish knot.

L—Lashing knot.

M—Trilashing knot.

N—Splice with thimble seized in.

O—Shackle with thimble spliced in close.

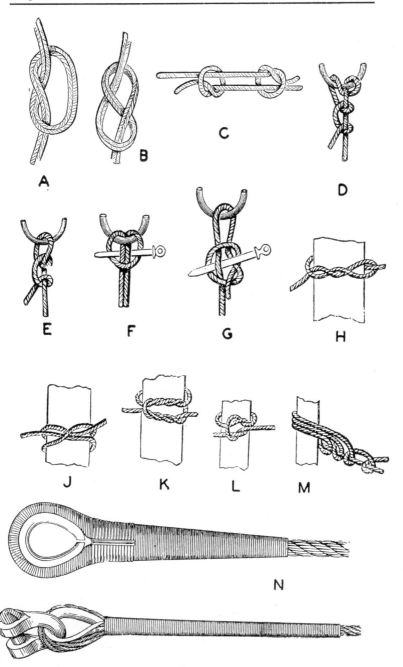

CLASS V.   ELEVATORS, DERRICKS, CRANES, CONVEYORS

Section 37k.   Knots, Hitches and Whipping

A—Clove hitch.
B—Slippery clove hitch.
C—Timber hitch.
D—Killick hitch.
E—Two half hitches.
F—Round turn and two half hitches.
G—Marlinespike hitch.
H—Blackwall hitch.
J—Double blackwall hitch.
K—Bill hitch.
L—Rolling hitch.
M—Stopper hitch.
N—Lifting hitch.
O—Lifting hitch with strap.
P—Sailor's square or reef knot.
Q—Reef point, slippery reef or draw knot.
R—Sheet bend or weaver's knot.
S—Reeving line bend.
T—Double lark's head.
U—Catspaw started.
V—Catspaw on hook.
W—Bowline.
X—Running bowline.
Y—Bowline on a bight.
Z—French bowline.
AA—Mousing a hook.
BB—Backsplice.
CC—Plain whipping started.
DD—Plain whipping finished.
EE—Wall knot shown open.
FF—Crown shown open.

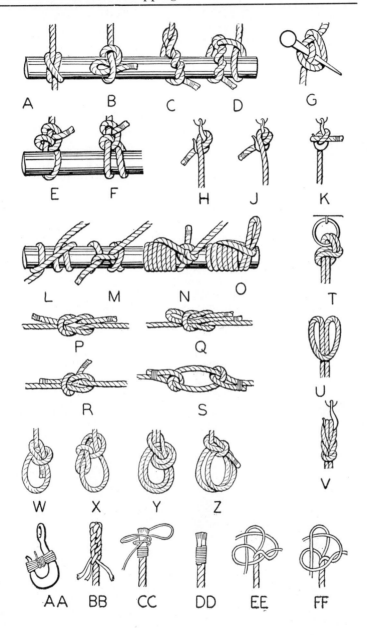

A   B   C   D   G

E   F   H   J   K

L   M   N   O   T

P   Q   U

R   S   V

W   X   Y   Z

AA   BB   CC   DD   EE   FF

CLASS V.   ELEVATORS, DERRICKS, CRANES, CONVEYORS

Section 371.   Rudimentary Feeders

A–Feeding gear for tickets.

B–Power or hand-feed gear for boring machines, drills, etc

C, D–Reversible feeding motion for shapers, etc.

E–Intermittent feeding gear; crank pin A strikes B and C alternately and there is a pause between each movement of the pawl.

F–Silent friction-pawl feed movement.

G–Pawl and spring in a sunk recess.

H–Kodak plate-feed apparatus.

J–Ink feeding for printing presses.

K–Feed worm with air blast.

L–Tube-rolling machine.

M–Drawing and throstle twisting rolls and bobbin winder; the front roll runs faster than the feed rolls and draws the fiber.

N–Cop winder; the arm and eye carry the thread forward and backward.

O–Cloth dresser; the central wheel is the teazel drum; the cloth is guided by rollers above and below.

P–Spool-winding machine; worm screw B and the gear drive a set of cams R and oscillate a lever and thread guide L back and forth. The spool spindle is driven by friction gear from shaft B.

Q–Railway sand box; the lever, pawl and ratchet wheel turn the worm feed.

R–Gas-pressure regulator; the gas flows in at the bottom and out at the side; and an inverted float is sealed in an annular cavity by mercury and is free to rise under excessive pressure and to close partially the inlet valve.

S–Creeper; an endless chain of boards, steel plate or buckets sliding along a wood or steel trough.

T–Vibrating electric feeder; it will feed any material that does not stick to the pan.

U–Reciprocating plate feeder.

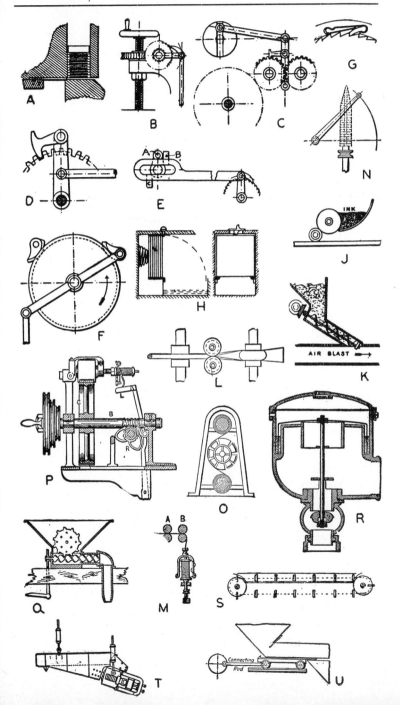

CLASS V.   ELEVATORS, DERRICKS, CRANES, CONVEYORS

Section 37m.   Commercial Feeders

A—Constant-weight vibrational feeder (Jeffrey).

B, C, D—Reciprocating feeder; average operating speed is fifteen strokes per minute; the stroke is adjustable to a maximum of 8 inches; the length of the pan should not exceed 8 feet; the capacity is regulated by the stroke and the depth of material is controlled by a gate in the hopper opening; B shows the feeder pan in the extreme backward position ready to start forward; C shows the feeder pan in extreme forward position; D shows the feeder pan during its backward motion to complete the cycle; this feeder can be adapted to handle sand, stone, ore, coal, coke, lime, etc. (Gifford-Wood).

E—Reciprocating feeder; shown beginning to advance (Link-Belt Co.).

F—Inclined apron feeder delivering coal to a single roll crusher placed over the lower run of a peck carrier (Link-Belt Co.).

G—Apron feeder delivering to a belt conveyor and driven by a motor and speed reducer (Link-Belt Co.).

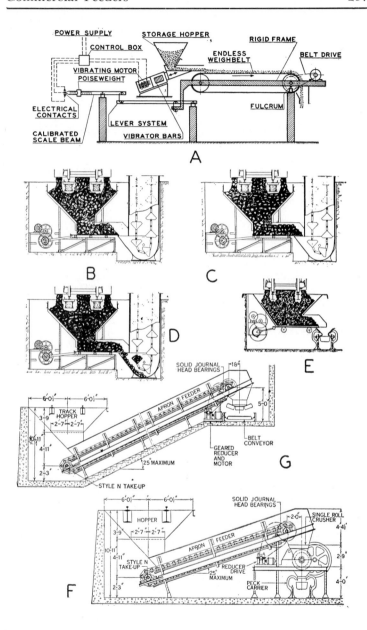

CLASS V.   ELEVATORS, DERRICKS, CRANES, CONVEYORS

Section 37n.   Bucket Elevators and Apron Conveyors

A—Elevator which at low speed drops the material instead of throwing it from the bucket (Caldwell).

B—Apron conveyor (Gifford-Wood Co.).

C—V-Bucket conveyor which combines the functions of both a bucket elevator and a scraper conveyor (Jeffrey).

D—Peck carrier for coal, ashes, gravel, stone, clinker, etc.; it consists of buckets, pivotally suspended between two endless chains; as the buckets maintain their carrying position by gravity, at all times, a single carrier can transport material horizontally, vertically and again horizontally, or in any desired path (within a vertical plane of travel); 36-inch pitch peck carrier is shown with 36-inch wide bucket (Link-Belt Co.).

E—Automatic power shovel, showing the box-car unloading hopper, screw feeder and bucket elevator to the storage bin; it has a hinged boom for the hopper and door snatch block (Link-Belt Co.).

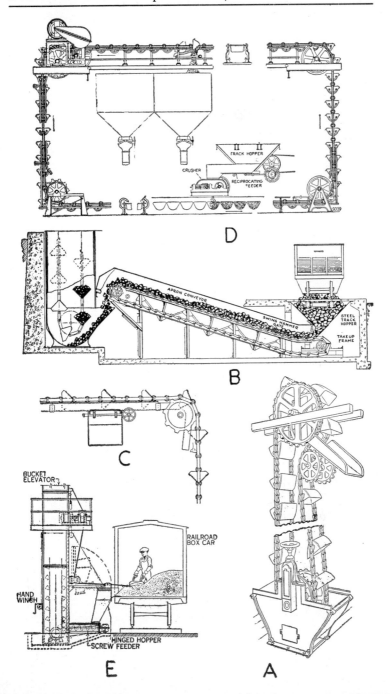

CLASS V.   ELEVATORS, DERRICKS, CRANES, CONVEYORS

Section 37o.   Bucket Elevators and Gravity-Discharge Conveyors

*A—Centrifugal discharge-type bucket elevator on chain or belt; loaded by material flowing into buckets, or, by their digging or scooping it up under the foot wheel; the material is discharged by centrifugal action as buckets pass over the head wheel.

B—Perfect discharge-type bucket elevator on double-strand chain; the buckets are carried between two chains, snubbed under head wheels to inverted position over the discharge chute; it operates successfully at low speeds for handling fragile, sticky, powdered or fluffy materials.

C—Continuous bucket elevator on single-strand chain or belt; spillage between buckets is prevented by their close spacing; and at the sides and front by a loading leg; the receiving chute is slightly narrower than the buckets and is so placed that there is always one bucket in the receiving position below the bottom of the receiving chute.

D—Super-capacity continuous bucket elevator on double-strand chain; the buckets and the operation are similar to those of C.

E—Inclined continuous bucket elevator for sand, gravel and stone, using medium front steel buckets on chain; it has no casing.

F to L—Six typical arrangements of gravity-discharge elevator conveyors with V-bucket carriers; used for elevating and conveying nonabrasive materials, e.g., coal, etc.

M—Pivoted bucket carriers are used for elevating and conveying material that will not stick to the buckets; they require less power than V-bucket carriers F to L, as the material is carried and not dragged on the horizontal run; the material is usually automatically fed on the lower horizontal run, elevated and discharged by a bucket-tripping device on the upper horizontal run.

N—Salem-type single-strand elevator back-hung buckets; used for short lifts handling free-flowing materials that will not twist the single strand.

O—Double-strand elevator end-hung buckets and pivoted attachments for heavier service.

P—Perfect-discharge elevator bucket as used for B.

Q—Continuous overlapping elevator buckets as used in C.

R—Detachable link.

S—Closed link with roller.

T—Closed link.

*A through T are Link-Belt Co. designs.

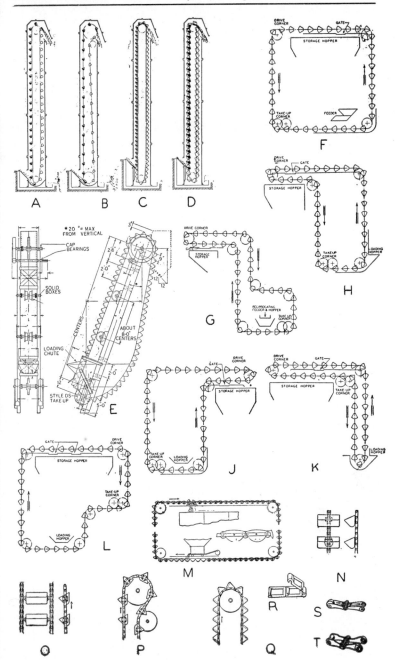

CLASS V.   ELEVATORS, DERRICKS, CRANES, CONVEYORS

Section 38a.   Hydraulic Elevators and Jacks

A –Typical arrangements of gravity-discharge elevator-conveyor.

B—Hydraulic jack of 20-ton capacity.

C—Balanced hydraulic elevator (Ellington).

D—Hydraulic elevator with pulley sheaves centered above the plunger.

E—Horizontal hydraulic elevator.

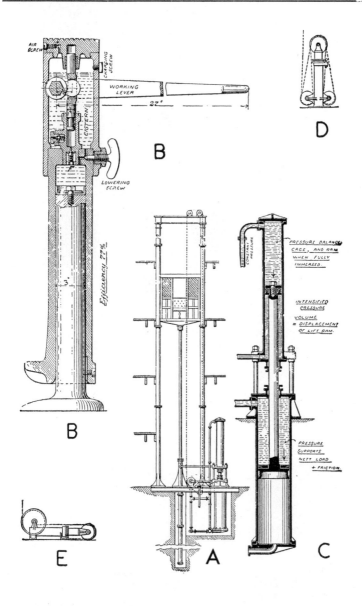

CLASS V.   ELEVATORS, DERRICKS, CRANES, CONVEYORS

Section 38b.   Balance Weights

A—Balanced cage of hoist.

B—Hydraulic balance elevator, in which the constant or dead load of the cage and ram is nearly balanced by a loaded piston in a supplementary cylinder; to raise the loaded cage, water pressure is admitted to the upper side of this piston.

C—Variable compensating balance for hydraulic-lift rams to compensate for loss from immersion of the ram as it ascends.

D—In deep lifts, to balance the weight, the chain or rope is made endless.

E—Another method of balancing in deep lifts is to have the loose chain hung from the cage of the same weight per foot as the lifting chain.

F—Increasing the balance by sections, lifted at intervals when the chain rises.

G—Variable volute-compensating balance for revolving shutters, curtains, blinds, etc.

H—Variable lever balance.

J—Compensating governor (Dawson).

K—Balanced doors hinged vertically.

L—Balanced sashes or vertical sliding doors or windows.

M—Method of balancing material in charging or withdrawing from a furnace, etc.

N—Weight to keep a rope in tension.

O—Method of balancing two sliding doors.

P—Balance for a suspended light.

Q—Balance box; the cover is as heavy as the box.

R—Balanced opening bridge.

S—Balanced rolling opening bridge.

T—Balanced bridge.

U—Counterbalance for two slides, which may be moved separately or together.

V, W—Compensating air cylinders employed in direct-acting horizontal pumps, working expansively, in lieu of a fly-wheel; the oscillating or vertical cylinders are air or spring pistons, absorbing power in the first part of the stroke and expanding during the second part (Worthington).

X—Dock crane; since the overhang is very great, the crane must be provided with a heavy frame and balance weight.

Y—Miscellaneous balance-weight designs for electric elevators.

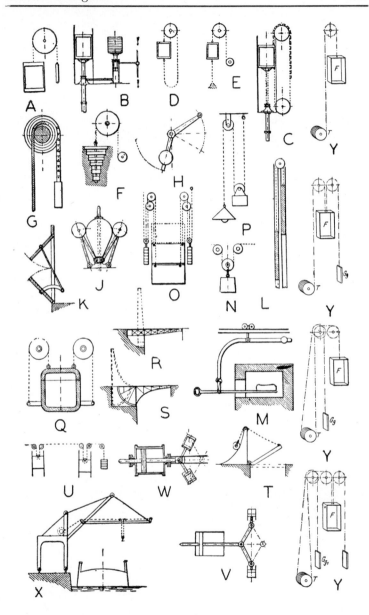

A B D E C Y

G F H P L Y

K J O N

Q R S M Y

U W T Y

X V Y

CLASS VI.   TRANSMISSION OF LIQUIDS AND GASES

Section 39a.   Primitive Water Lifts*

A—Persian wheel driven by current; the water follows the curved arms and discharges through the hollow shaft while the buckets, suspended at the periphery, are tipped at top into a trough.

B—Archimedes-screw water lift.

C—Earthern pots fastened to the rim of a revolving wheel actuated by the stream of water.

D—Drainage wheel; it is power-driven and lifts a large volume of water to a height of nearly half its diameter.

E—Water-raising current wheel with bucket.

F—Fixed-bucket water-raising current wheel.

G—Chain pump.

H—Teeter pump (obsolete design).

J—Well pulley and bucket.

K—New England sweep.

L—Fairburn bailing scoop.

M—Water-raising current lift.

N—Pendulum water lift having scoops with flap valves and connecting pipes; swinging of the pendulum frame alternately immerses the lower scoops and raises the water to the opposite scoop, etc.

O—Bellows pump or blower (ancient).

*Some of these are still in use.

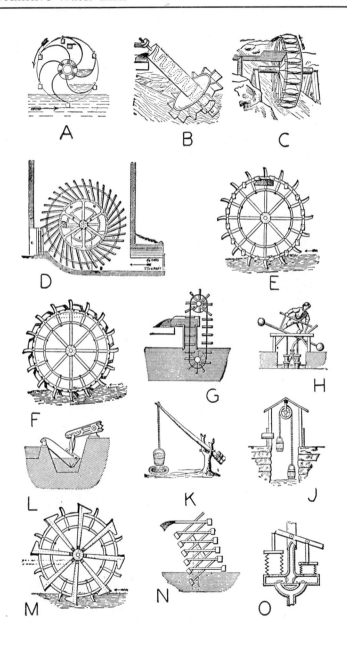

A

B

C

D

E

F

G

H

L

K

J

M

N

O

CLASS VI.   TRANSMISSION OF LIQUIDS AND GASES

Section 39b.   Rudimentary Pumps

A—Gravity motor which acts as a pump when reversed; the floats are placed about 10 feet apart; the slip is about 20 per cent; the chain speed varies from 200 to 300 feet per minute; it is suitable for lifts up to 60 feet; its efficiency is about 63 per cent. (The ordinary dredger is a pump of this type.)

B—Reversed undershot or breast wheel used for low lifts; the circumferential speed is 6 to 10 feet per second; the slip varies from 5 to 20 per cent; the diameter of the wheel varies from 20 to 50 feet; the width of the paddle varies from 1 to 5 feet; the pitch of paddles is about 18 inches; its efficiency varies from 50 to 70 per cent.

C—Bucket-type single-acting pump; it gives an intermittent discharge and is suitable for low lifts.

D—Double-acting forced or piston-type reciprocating pump suitable for high lifts; when made single-acting, the upper set of valves is eliminated.

E—Single-acting plunger pump; adopted for very high pressures.

F—D'Auria's pendulum arrangement on direct-acting steam pumps to overcome nonexpansion of the steam.

G—D'Auria's water compensator; built on the same principle as F.

H—Impulse ram.

J—Hand suction pump.

K—Force pump.

L—Jet pump.

M—Diaphragm pump; its flexible diaphragm functions as a piston.

N—Parallel motion for piston pumps.

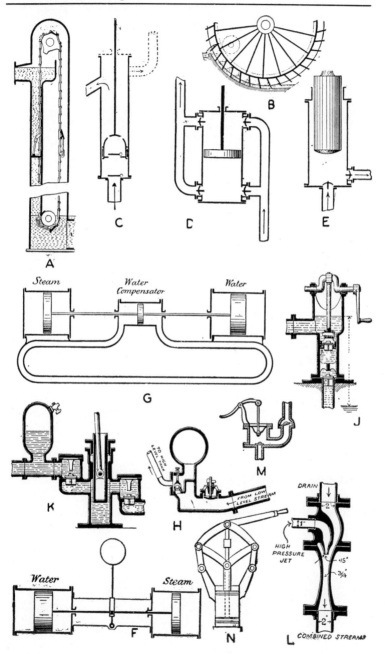

CLASS VI.   TRANSMISSION OF LIQUIDS AND GASES

Section 39c.   Commercial Reciprocating Pump Classes

A—Direct-acting pump; liquid submerged; packed piston; double acting; valve plate; disc valves.

B—Liquid end, straight-way-packed piston pump; double acting; disc valves.

C—Liquid-end, center-packed-plunger, double-acting pump; submerged; valve plate; disc valves.

D—Direct-acting pump; liquid submerged; packed piston, double acting; valve plate; disc valves.

E—Liquid-end, straight-way-packed piston pump; double acting; disc valves.

F—Liquid-end, center-packed-plunger, double-acting pump; submerged; valve plate; disc valves.

G—Direct-acting pump; liquid submerged; packed piston; double acting; valve plate; disc valves.

H—Liquid-end, straight-way-packed piston pump; double acting; disc valves.

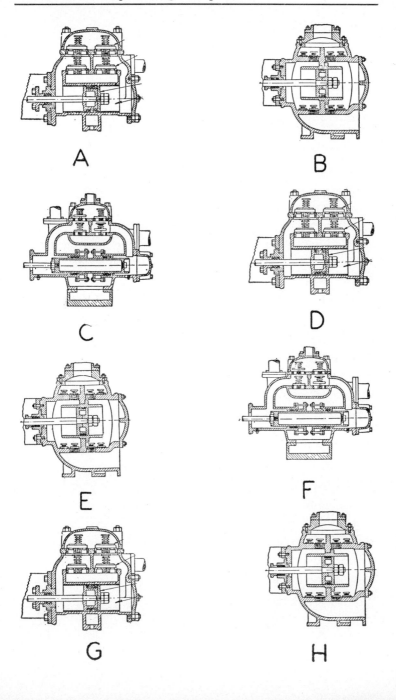

A

B

C

D

E

F

G

H

CLASS VI.   TRANSMISSION OF LIQUIDS AND GASES

Section 39d.   Reciprocating-Pump Valves

A—Rubber disc and grating for reciprocating-pump valve.

B—Multiple rubber-ball valve for high-lift discharge valves of large pumping engines.

C—Multiple-ring or disc valve; the rings open and close in succession, thus avoiding shocks.

D—Double-beat cornish or equilibrium valve.

E—Double-beat valve with sunk seating; it may be loaded by spring or weight and used as a relief valve.

F—Double-beat ring valve.

G—Simple flap valve faced with leather or rubber.

H—Rocking or rolling valve for closing and opening easily and gradually against pressure.

J—Roll-up valve for the same purpose as H.

K—West's spiral valve with rubber cord; it expands and contracts over spiral perforated grooves.

L—Compound-flap valve.

M—Four-seated sunk valve.

N—Dished-grating valve.

O—Rubber pump valve.

P—Double-flap valve.

Q—Spring-flap valve.

R—Double-flap grating valve.

S—Multiple-ring valve; the rubber band expands and contracts over perforations.

T—Cup valve and suspended weight.

U—Oscillating valve.

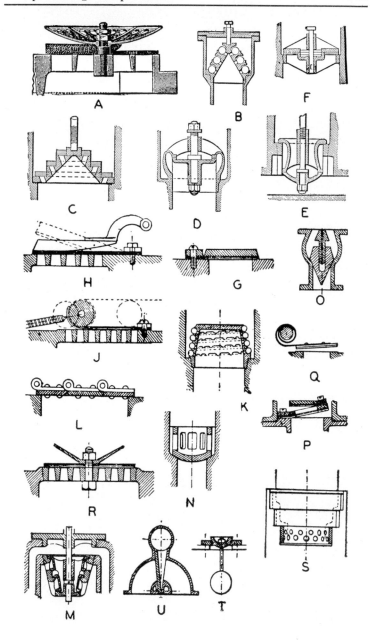

CLASS VI. TRANSMISSION OF LIQUIDS AND GASES

Section 39e. Modern Commercial Pump Valves

A—Pot-wing valve.

B—Rubber-disk valve for cold water (Myers).

C—Plunger with hemp packing for handling oil, kerosene or gasoline.

D—Brass valve for handling hot water, kerosene and gasoline.

E—Plunger for hot-water service with hydraulic packing.

F—Dual-seated valve with guides for a slush-pump.

G—Disk valve; suction and discharge on the same stem.

H—Bowl valves for heavy viscous liquids.

J—Ball valves of glass, brass, steel or rubber for thick liquids.

K—Leather-disk-type valve (Weise and Monski, Germany).

L—Schroder valve (German).

M—Valve by Maschinenfabrik Oddesse (German).

N—Valve by A. Borsig (German).

O—Valve by Balcke (German).

P—Valve by Koerting (German).

Q—Valve by A. Borsig (German).

R—Valve by A. Borsig (German).

S—Condenser air pump without suction valves.

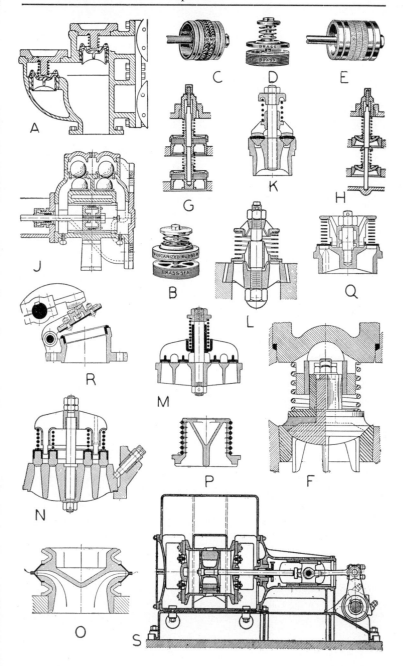

CLASS VI.   TRANSMISSION OF LIQUIDS AND GASES

Section 40a.   Leather Packings

A–Leather collar in a groove.

B–Leather collar bent into a crescent shape for insertion into a groove.

C–Leather collar with gland; when water under pressure enters space S the hole in the collar is made slightly smaller than the ram to give a close fit, and the outer part is slightly larger in diameter than the groove for the same reason; the pressure forces one part of the collar against the ram and the other side against the groove.

D–Worn-out leather collar showing that the greatest wear occurs at the shoulder B of figure C.

E–Piston fitted with double leather cup for hydraulic elevator or crane.

F–Press for molding cup leathers.

G–Cylinder of a hydraulic capstan (section).

H–Packing for small plunger pump.

J–Finger-type expander to keep the walls of the packing in contact with the cylinder. (Graton and Knight Co.).

K–Coil-spring expander.

L–Face of gland showing correct design with flat gland face, bull ring and flat hemp or leather filler.

M–Face of gland showing incorrect design with curved gland face.

N–Vee leather packing showing the assembly before pressure is applied.*

O–Vee leather packing showing the assembly under pressure.

P–Leather-backing washers on top of a packing will prevent the nut or gland from cutting the flanged leather; coiled spring creates permanent seal (Graton and Knight Co.).

Q, R–V-ring-type hydraulic packing (Linear, Inc.).

S–Critical dimensions of V-ring hydraulic packing (Linear, Inc.).

---

*U-leather packing is generally used to seal a reciprocating shaft or piston (seldom used with a rotating shaft); Vee-leather packings may be used with a rotating shaft.

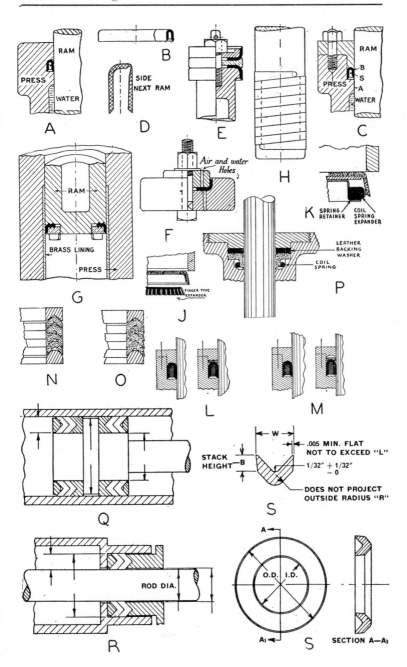

A

B

SIDE NEXT RAM

D

E

H

C

RAM

PRESS

WATER

B

S

A

G

RAM

BRASS LINING

PRESS

F

Air and water Holes

K

SPRING RETAINER   COIL SPRING EXPANDER

LEATHER BACKING WASHER

COIL SPRING

P

J

FINGER TYPE EXPANDER

N

O

L

M

Q

S

W

.005 MIN. FLAT NOT TO EXCEED "L"

STACK HEIGHT   B

1/32" + 1/32" − 0

DOES NOT PROJECT OUTSIDE RADIUS "R"

R

ROD DIA.

A

O.D.   I.D.

A₁

S

SECTION A—A₁

CLASS VI.   TRANSMISSION OF LIQUIDS AND GASES

Section 40b.   Metallic Packings

A—Normal position of France patented ring (France Packing Co.).

B—France ring extended showing the parallel movement of joints.

C—Inside split packing containing three pairs of wearing rings suitable for pressures up to 150 psi; with four pairs, up 175 psi.

D—Inside annular packing with three pairs of wearing rings.

E—Outside annular packing.

F—Outside split packing.

G—Split-flange packing.

H—Oil-return packing for engine baffle plate.

J—Rings and casing furnished for engine baffle plate.

K—Annular-flange packing.

L—Annular-flange packing for gas compressors; the number of wearing rings depends on the amount of pressure to be held; it can be provided with a vent connection in addition to the lubrication connection (France Packing Co.).

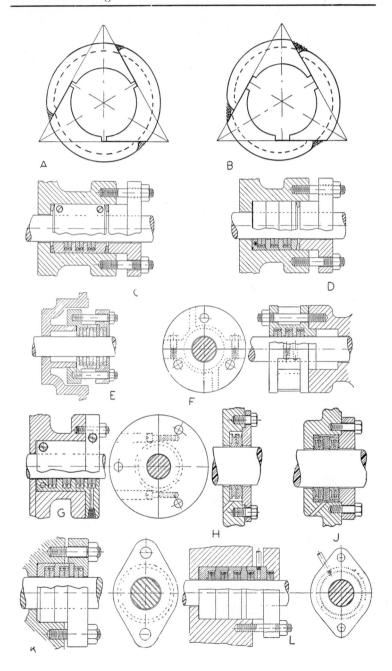

CLASS VI.   TRANSMISSION OF LIQUIDS AND GASES

Section 40c.   Pistons

A—Steam-engine piston with snap rings.

B—Bucket pump.

C—Plunger pump.

D—Pump piston with packing rings.

E—Grooved solid-metal piston; used in steam-engine indicators, etc.

F to K—Snap rings sprung over the piston; hard to remove on large cylinders.

L, M—Junk ring; it permits easy removal of rings.

O, P, Q—Piston with junk-ring part.

R—Oil-engine piston with auxiliary internal snap-ring cast-iron spring.

S, T, U—Marine piston with auxiliary flattened helical spring.

V—Cameron corrugated spring; it does not touch the body of the piston.

W—Early-type coach springs.

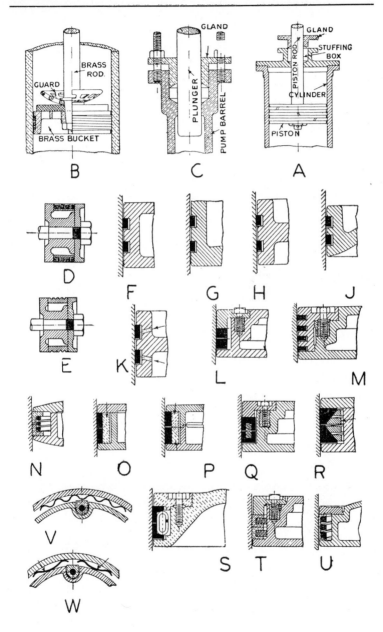

CLASS VI.   TRANSMISSION OF LIQUIDS AND GASES

Section 41a.   Early-Type Rotary Pumps

A—Pappenheim rotary pump.

B—Cochrane rotary pump.

C—Rotary pump or motor; can be run in either direction.

D—Cary rotary pump.

E—Pattison rotary pump.

F—Ramelli rotary pump.

G—Emery rotary pump.

H—Heppel rotary pump.

J—Knott rotary pump.

K—Repsol rotary pump.

L—Holley rotary pump.

M—Quimby screw pump.

N—Root's blower and pump.

O—Root's blower.

P—Mackenzie's blower; it may have one, two or three vanes.

Q—Gould's rotary pump.

R—Greindl rotary pump.

S—Mellory rotary pump; consists of a rocking vane or partition with packing device which accommodates itself to the revolving oval piston.

T—Rotary motor or pump; it has four rolling pistons.

U—Star-wheel-geared rotary pump or motor.

V—Hasafan eccentric revolving piston.

W—Two hinged vanes and eccentric rotor or piston.

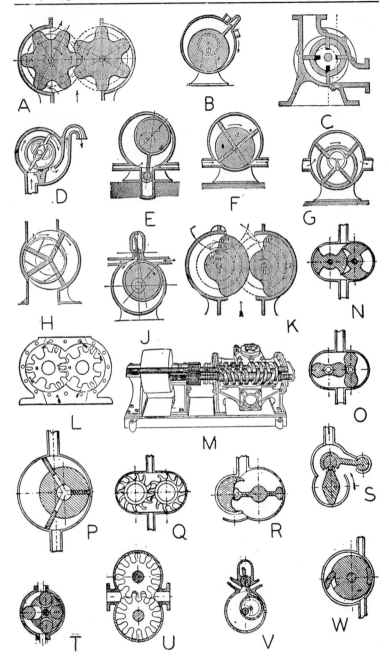

CLASS VI.   TRANSMISSION OF LIQUIDS AND GASES

Section 41b.   Commercial Rotary Pumps

A—Blackmer swinging-vane-type rotary pump; made in ten sizes from 10 to 750 gpm; operating pressures are from 5-inch mercury gage vacuum to about 60 psi.

B—Blackmer relief-valve design.

C—Blackmer strainer.

D—Northern internal gear pump; a driving pinion meshes with and drives a larger internal gear, which fits snugly in the pump housing; a stationary crescent, which is part of the housing, fills the space that is not swept by the moving teeth.

E—Gerotor pump; a special form of internal-gear pump; an inner gear is keyed to, and rotates with, the driving shaft; an outer gear of internal type is driven by the inner gear and is free to rotate with a snug fit in a recess in one end of the housing; the teeth of the two gears are specially shaped so that the tops of all teeth of the inner gear are always in sliding contact with the teeth of the outer gear.

F—Vickers vane pump; a constant-discharge pump in which radial vanes produce the pumping action; the vanes are free to slide in and out of a rotating hub and so maintain contact with the outer ring; oilways from the high-pressure side of the pump to the spaces behind the vanes assure that this contact is maintained at all times.

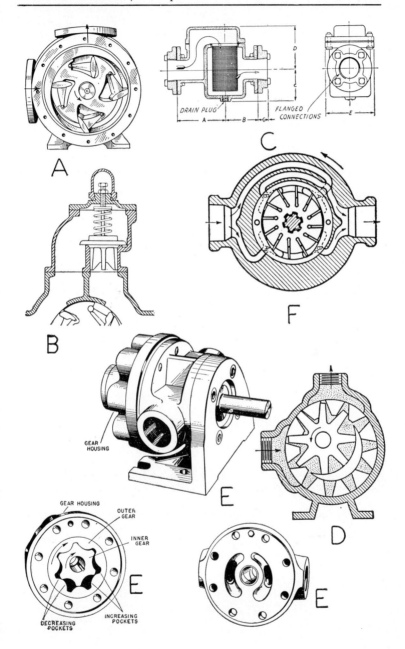

DRAIN PLUG

FLANGED CONNECTIONS

A

B

C

F

GEAR HOUSING

E

D

GEAR HOUSING   OUTER GEAR

INNER GEAR

E

DECREASING POCKETS   INCREASING POCKETS

E

CLASS VI.   TRANSMISSION OF LIQUIDS AND GASES

Section 41c.   Commercial Rotary Pumps

A—Cylinder block of West axial hydraulic pump; it is rigidly connected to the driving shaft and is rotated in a stationary housing.

B—Piston element of the West axial hydraulic pump which rotates with the cylinder block.

C, D, E—Assemblies of A and B; this pump is of the variable-discharge reversible-delivery type.

F—Racine vane pump; a constant-pressure pump which permits variable discharge.

G—Hele-Shaw radial-piston pump; neutral position.

H—Same as G, but the piston assembly has been rotated and moved to the left.

J—Same as G, with the piston assembly rotated to the left.

K—Screw-type rotary pump for pumping heavy oils.

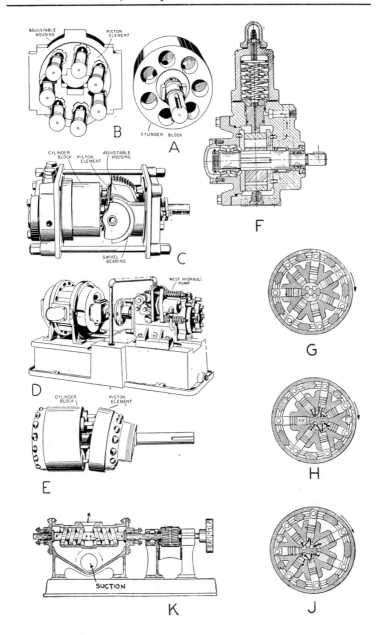

CLASS VI.   TRANSMISSION OF LIQUIDS AND GASES

Section 41d.   Commercial Rotary Pumps

A, B, C—Hele-Shaw radial-piston pump; sliding shoes maintain contact between the pistons and the outer ring; drilled passages in the stationary spindle lead to internal parts that register with the open ends of the revolving cylinders.

D—Oilgear radial-piston pump. Two different designs have been developed for pressures of several thousand pounds per square inch; the one illustrated employs rollers to make contact between the pistons and the outer ring as shown in G.

E, F, G—Northern radial-piston pump; it employs sliding shoes to make contact between the pistons and the outer ring.

H, J, K—Northern radial-piston pump employing pistons moved in and out by rollers which make contact with the outer ring.

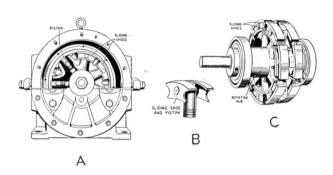

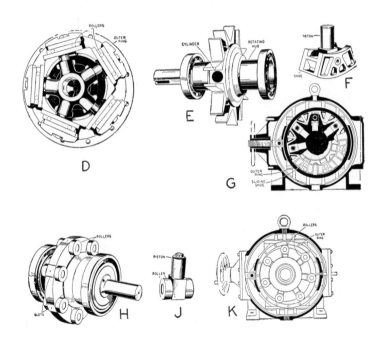

CLASS VI.   TRANSMISSION OF LIQUIDS AND GASES

Section 41e.   Typical Rotary Pumps

A—External-screw pump.

B—Steam-jacketed cam and piston pump.

C—Spur or external-gear pump.

D—External-screw pump.

E—Two-lobe pump.

F—Internal-gear pump.

G—Helical-gear pump.

H—Sliding-vane pump.

J—Three-lobe pump.

K—Internal-screw pump.

L—Swinging-vane pump.

M—Special-vane pump.

N—Internal-gear pump.

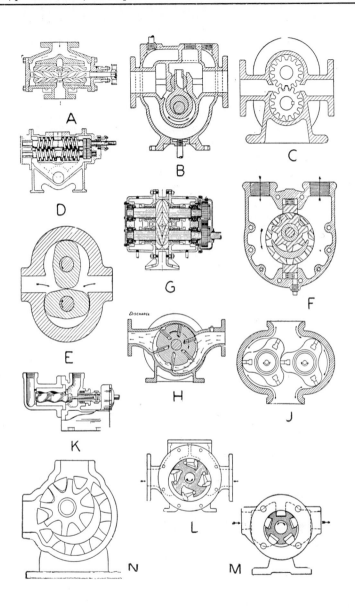

A

B

C

D

G

F

E

H

J

K

L

N

M

CLASS VI.   TRANSMISSION OF LIQUIDS AND GASES

Section 42a.   Early-Type Air Blowers

A—Three-throw bellows; the power is supplied by a crank; it gives constant-volume blast without an equalizer.

B—Double-organ blowing bellows; the upper compartment equalizes the alternating air pressure from the two blower sections.

C—Foot bellows for a blow pipe; the netting prevents rupture of the rubber bag.

D—Root early design of a rotary blower.

E—Root later design of a rotary blower.

F—Fabry rotary blower.

G—Wedding rotary blower.

H—Old-style forge blower.

J—Disstons rotary blower.

K—Bagley and Sewall's rotary blower.

L, M, N, O—Varieties of intergeared rotary blowers.

P—Rotary blower; a crescent-shaped piston rotates three times to one revolution of the three-winged piston.

Q—Rotary blower; a hinged shutter is cleared out of the way each time the revolving arm passes it; rotary blower design.

R—Rotary blower with sliding shutter and cam device.

S, T—Miscellaneous Root blowers.

U—The hinged vanes close on the revolving piston when passing the flat side of the casing.

V—Eccentric piston and sliding diaphragm.

W—Klein's rotary blower motion.

X—Baker's pressure blower.

Y—Rotary blower with eccentric piston and sliding diaphragm.

Z—Rotary blower with eccentric piston and two sliding vanes.

AA—Mellor's rotary blower; it consists of a rocking vane oscillated by an eccentric piston.

BB—An eccentric four-armed piston with four rolling stoppers actuated by centrifugal force.

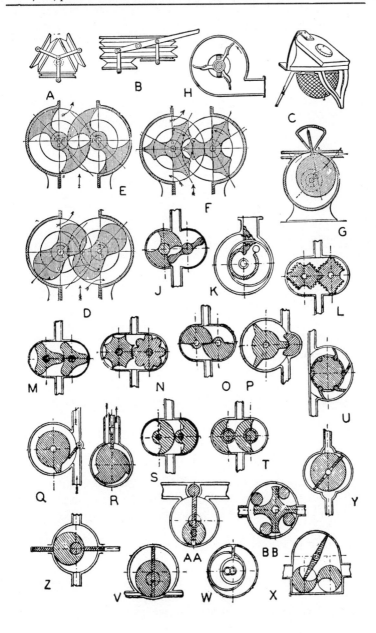

CLASS VI.   TRANSMISSION OF LIQUIDS AND GASES

Section 42b.   Commercial Blowers and Compressors

A—Rotary compressor used to compress gas to about 75 psi.

B—Root's supercharger; positive-pressure type.

C—Pendulum compressor; positive-pressure type.

D—Vane-type positive-pressure blower.

E—Norge refrigerating compressor showing the cylinder full of gas at the start of compression.

F—Norge refrigerating compressor, showing the discharge valve open on compression stroke.

G—Norge refrigerating compressor, showing the discharge and suction strokes completed.

H—Typical single-stage centrifugal blower; used on oil burners.

J—Pressure blower; paddle-wheel type; planing-mill exhauster.

K—Centrifugal-type fan; multiblade, nonoverloading type; used in ventilation and air-conditioning applications for silent operation and large volumes of air.

L—Conical-plate fan.

M—Cooke's ventilator; early English type, in first position.

N—Cooke's ventilator, in second position.

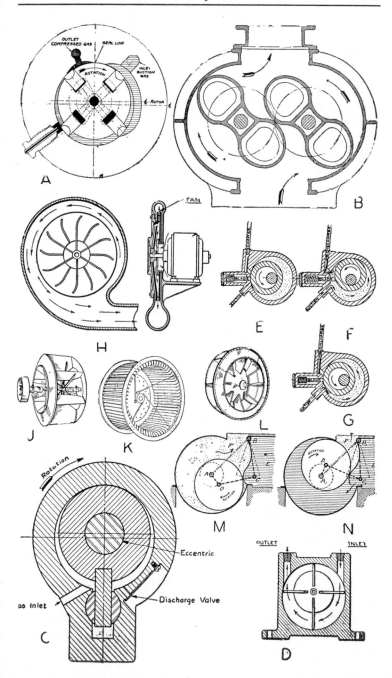

CLASS VI.   TRANSMISSION OF LIQUIDS AND GASES

Section 42c.   Small Refrigerating-Compressor Valves

A, B, C—Early-type discharge valves for small refrigerating units.

D, E, F—Early-type suction valves for small refrigerating units.

G—Discharge-valve assembly; reed type.

H—Suction valve; reed type.

J—Suction valve in the piston.

K—Discharge-valve design.

L—Single-acting, encased crank refrigerating compressor, showing the suction valve in the piston, equalizing port and discharge valve.

M—Suction-valve assembly.

N—Discharge-valve assembly.

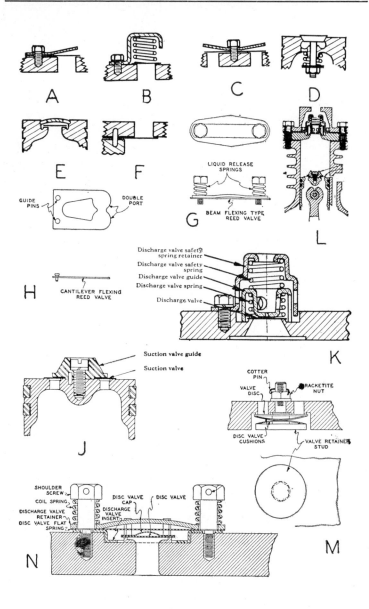

A B C D

E F

GUIDE PINS   DOUBLE PORT

LIQUID RELEASE SPRINGS

BEAM FLEXING TYPE REED VALVE
G

L

Discharge valve safety spring retainer
Discharge valve safety spring
Discharge valve guide
Discharge valve spring
Discharge valve

H   CANTILEVER FLEXING REED VALVE

K

Suction valve guide
Suction valve

COTTER PIN
VALVE DISC
RACKETITE NUT

DISC VALVE CUSHIONS
VALVE RETAINER STUD

J

M

SHOULDER SCREW
COIL SPRING
DISCHARGE VALVE RETAINER
DISC VALVE FLAT SPRING
DISC VALVE CAP   DISC VALVE
DISCHARGE VALVE INSERT

N

CLASS VI.   TRANSMISSION OF LIQUIDS AND GASES

Section 42d.   Refrigerating Compressors and Valves

A—Refrigeration compressor classified by action and position; vertical single-acting.

B—Refrigeration compressor classified by action and position; horizontal double-acting.

C—Refrigeration compressor classified by suction-valve location; in piston.

D—Refrigeration compressor classified by suction-valve location; in cylinder head.

E—Refrigeration compressor classified by suction-valve location; in cylinder wall.

F—Refrigeration compressor classified by stages; single-stage.

G—Refrigeration compressor classified by stages; two-stage.

H, J—Plate-valve design for large refrigeration compressors.

K, L—Improved plate-valve design for large refrigeration compressors.

M—Section through an assembled Ingersoll-Rand plate valve.

N, O, P, Q, R—Typical German compressor-valve designs.

S—Comparison of a Voss discharge-valve plate with the old-style poppet valve.

T—Old-style poppet valve.

U—Voss ring-plate valve.

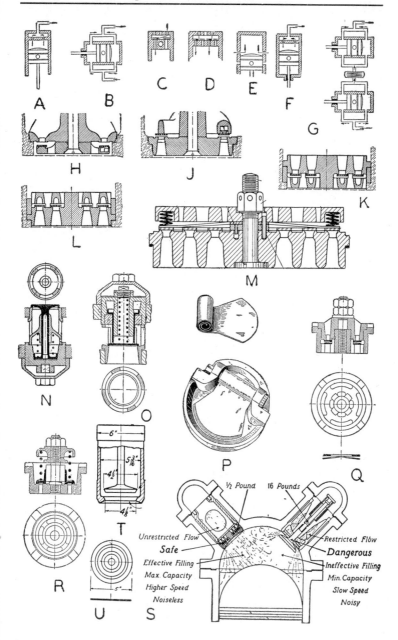

CLASS VI.   TRANSMISSION OF LIQUIDS AND GASES

Section 42e.   Blowing Engine and Valves

A to G—Blowing cylinder showing grid valves.

H—Plate valve.

J—Mechanically operated air-compressor valve.

K—Poppet-type discharge valve.

L—Corliss-type air valve.

M—Horizontal blowing engine driven by a cross compound steam engine.

N—Horizontal blowing cylinder direct driven by a gas engine.

O—Valve seat.

P—Valve retainer.

Q—Slotted plate.

R—Springs

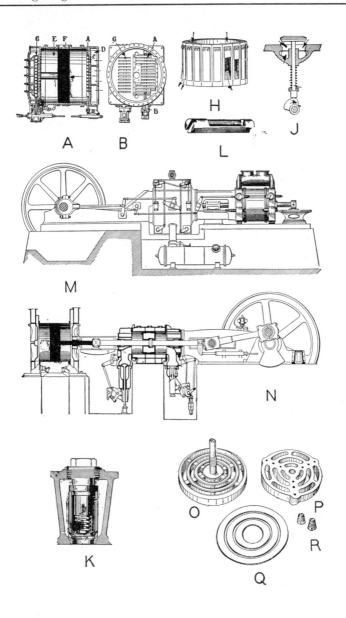

A B

H

L

J

M

N

K O P R Q

CLASS VI.   TRANSMISSION OF LIQUIDS AND GASES

Section 43a.   Refrigeration-Fluid Control

A—Automatic expansion valve for small commercial refrigeration systems using Freon.

B—Alco thermostatic expansion valve.

C—Diaphragm-type automatic direct-expansion valve.

D—Low side-float liquid-control evaporator and ebullator.

E—High-pressure float valve for refrigerant-liquid control.

F—Magnetic liquid shut-off valve.

G—Safety valve used on ammonia systems.

H—Back-pressure valve for ammonia systems.

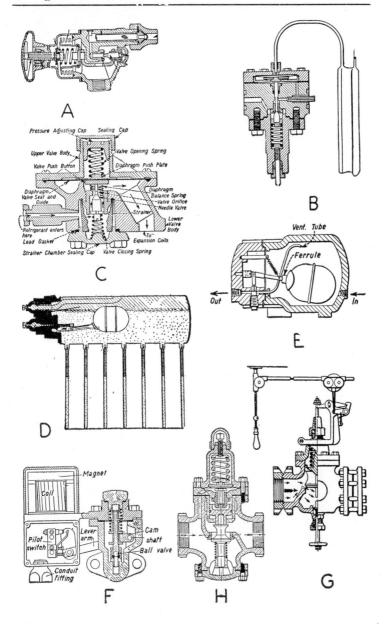

A

C

Pressure Adjusting Cap   Sealing Cap

Upper Valve Body   Valve Opening Spring

Valve Push Button   Diaphragm Push Plate

Diaphragm
Valve Seat and
Guide   Diaphragm
                 Balance Spring
                 Valve Orifice
                 Needle Valve
                 Strainer
                 Lower
                 Valve
                 Body
Refrigerant enters
here   To
Lead Gasket   Expansion Coils

Strainer Chamber Sealing Cap   Valve Closing Spring

B

E

Vent. Tube

Ferrule

Out   In

D

F

Magnet

Coil

Pilot
switch

Lever
arm

Cam
shaft

Ball valve

Conduit
fitting

H

G

CLASS VI.   TRANSMISSION OF LIQUIDS AND GASES

Section 43b.   General-Purpose Shut-Off Valves

A—Elliott double-automatic boiler header valve shown open.

B—Elliott double-automatic boiler header valve shown closed.

C—City water-pressure regulating valve with sylphon bellows.

D—Packless refrigeration shut-off valve with sylphon bellows.

E—Globe valve (Schaffer and Budenberg, Germany).

F—Globe angle valve (German).

G—High-pressure steam valve (Ferranti patent).

H—Cock (German).

J—Pressure-reducing valve (Hubner and Mayer, Austria).

K—Pressure-reducing valve (Hubner and Mayer).

L—Pressure-reducing valve (German).

M—Gate valve (German).

N to T—Typical valve-stem stuffing boxes (German).

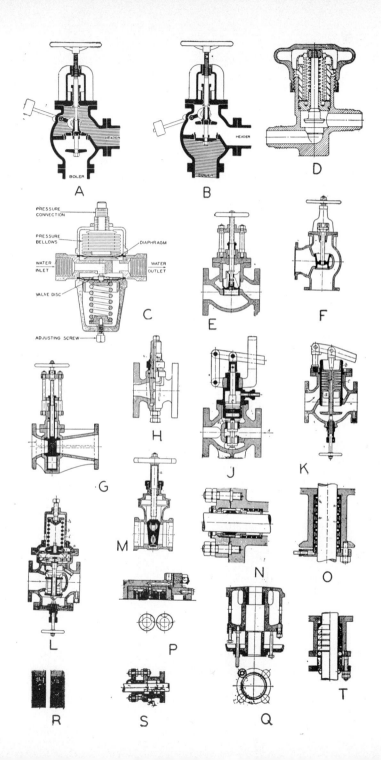

A

B

D

PRESSURE
CONNECTION

PRESSURE
BELLOWS   DIAPHRAGM

WATER
INLET   WATER
                                              OUTLET

VALVE DISC

ADJUSTING SCREW

C

E

F

G

H

J

K

L

M

N

O

P

R

S

Q

T

CLASS VI.   TRANSMISSION OF LIQUIDS AND GASES

Section 43c.   Lubrication Methods

A—Beauchamp Tower's experiment (1885), showing the variation of oil pressure which is highest on the off side.

B—Oil hole drilled at the point of highest pressure; may be forced out at certain speeds.

C—Correct position of oil holes at the point of minimum pressure for better lubrication.

D—Needle method of fluid lubrication; drop feed with sight-feed oilers.

E—Syphon method of fluid lubrication.

F—Pad method of fluid lubrication.

G—Bath method of fluid lubrication; the friction is the same as with forced feed.

H—Ring method of fluid lubrication.

J—Collar bearing of fluid lubrication.

K—Pivot or footstep of fluid lubrication.

L—Screw-top oil cup; gravity feed.

M—Screw-top oil cup; wick feed.

N—Sight gravity-feed oiler with needle-valve shut-off and breather tube.

O—Gits patented wick-feed multiple oiler with flushing features.

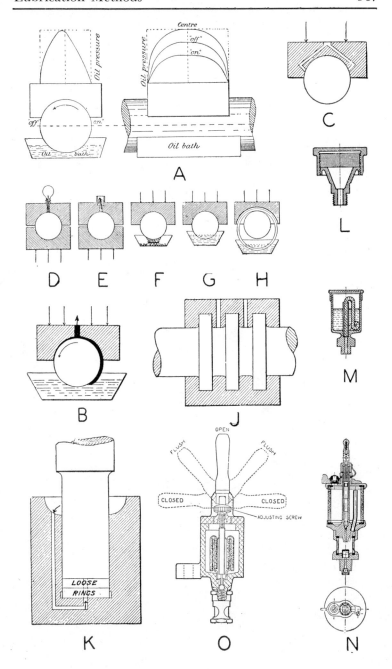

A

C

L

D E F G H

M

B

J

K O N

CLASS VI.   TRANSMISSION OF LIQUIDS AND GASES

Section 43d.   Nozzles and Jets

A—Rose jet for spreading.

B—Short jet.

C—Straight jet for greater distance.

D—Fan jet.

E—Blast tuyere.

F—Injector for fuel oil and air, or stream and air blast.

G—Smith's tuyere.

H, J--Spray jets; aspirator.

K—Spreading jet; the vanes are inserted into the jet of water to cut it up.

L—Ventilating jet or aspirator with several openings for inducing a current.

M—Water-jet condenser.

N—Spray nozzle with spiral core.

O—Spray jet with air blast.

P—Spray jet with annual orifice and dash plate.

Q—Revolving jet.

R—Ring spray jet.

S—Steam jet for water heating.

T—Hollow jet for spraying.

U—Jet aspirator for inducing a mixed current of water and air or steam.

V—Automatic sprinkler head; at about 165°F, the solder melts, the link separates and the water pressure opens the valve, thus supplying a spray to extinguish a local fire.

W, X—Typical mechanical pressure-atomizer tips for oil burners.

Y—Low-pressure fan-air atomizer for oil burners.

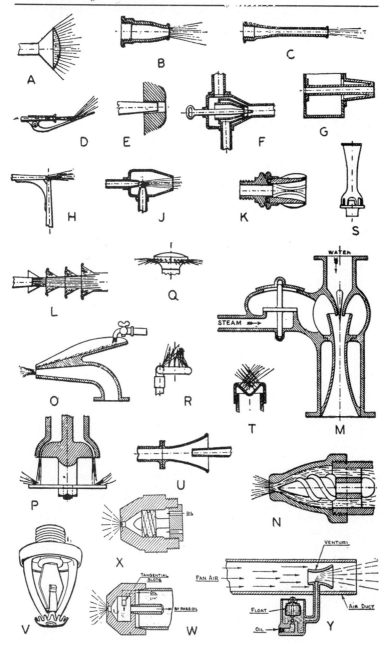

CLASS VI.   TRANSMISSION OF LIQUIDS AND GASES

Section 43e.   Steam Traps and Separators*

A–Condensation return steam trap; located above the water line of a boiler (early Blessing type).

B–Spring steam trap; a differential expansion of a spring, made of two metal strips —the upper one of brass and the lower one of steel— riveted together, causes the valve to open at water temperature and close at steam temperature.

C–Bucket steam-trap; water condensed in a heating system flows into the trap, closing the valve by raising the bucket; overflow into the bucket causes it to sink, opening the valve and discharging the water.

D–Bundy steam trap; the bowl rises when empty and falls when filled with water; it swings on trunnions carrying an arm which opens and closes a valve.

E–Balanced-float steam trap with trial handle.

F–Armstrong inverted-bucket steam trap.

G–Armstrong "snap-action" air trap; draining water from compressed air lines presents a difficult problem; the high cost of compressed air makes a leak more expensive; a snap-action trap is, therefore, desirable; the ball float is connected to the short valve lever through a flat strip of stainless spring steel; in the closed position, this spring is bowed downward, as water enters the body of the trap, the ball float rises and stores up energy in the spring; just before the ball float reaches the top of the trap, the spring bends past dead center and the stored-up energy snaps the valve wide open; in this position, the spring is bowed upward; as the water level drops in the trap body, the cycle is reversed and the valve snaps shut.

H–Inverted bucket trap; for pressures up to 250 psi (Watson and McDaniel Co., Philadelphia, Pa.).

J, K–German-type bucket traps.

L–Steam separator.

M–Austin steam separator.

*See section 68 for steam hook-ups.

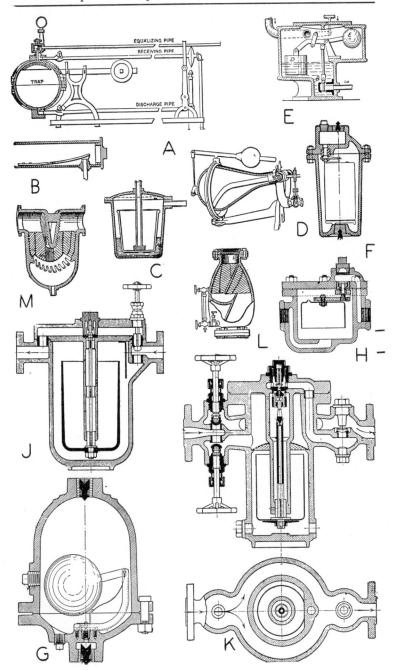

A

B

C

D

E

F

G

H

J

K

L

M

CLASS VII. COMBUSTION

Section 44a. Early Boiler Types

A—Boiler with vertical return flues.

B—Field boiler with internal circulating tubes and suspended tubes.

C—Vertical multitube boiler.

D—Center-flue boiler.

E—Vertical egg-end boiler, with spiral flue.

F—Vertical boiler with diagonal tubes and smoke box.

G—Pot boiler.

H—Horizontal-return tubular boiler.

J—Locomotive-type boiler with the fire box underneath.

K—Locomotive multitubular boiler; easily cleaned.

L—Egg-end boiler; was used where fuel was very cheap.

M—Multitubular, horizontal, self-contained boiler.

N—Cornish single-flue boiler with enlarged fire-box tube.

O—Lancashire double-flue boiler.

P, Q—Elephant boilers; coke-oven and flue-gas applications.

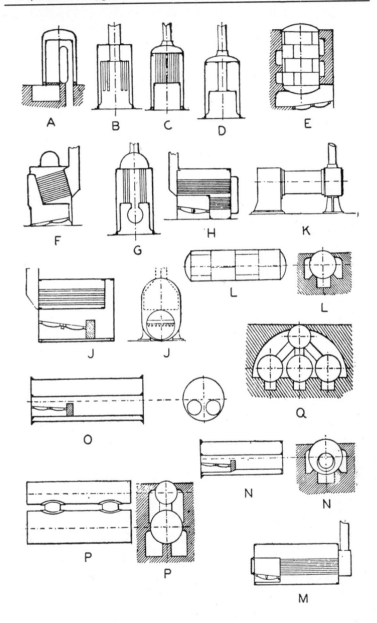

CLASS VII.   COMBUSTION

Section 44b.   Early Coal-Fired Boilers

A—Cylindrical boiler with hanging water drum.

B—Plain cylindrical boiler.

C—Cylindrical double-flue boiler.

D—Stevens boiler; very early type from about 1810.

E—Dion steamer automobile boiler.

F—Thornecroft water-tube boiler.

G—See water-tube boiler.

H—Yarron water-tube boiler.

 J—Boyer's water-tube boiler.

K—Hazleton boiler.

 L—Moyes water-tube boiler.

M—Cahall water-tube boiler.

N—Scotch marine boiler with steel combustion chamber.

O—Herreshoff boiler.

 P—Niclausse's water-tube boiler.

Q—Babcock and Wilcox's water-tube boiler.

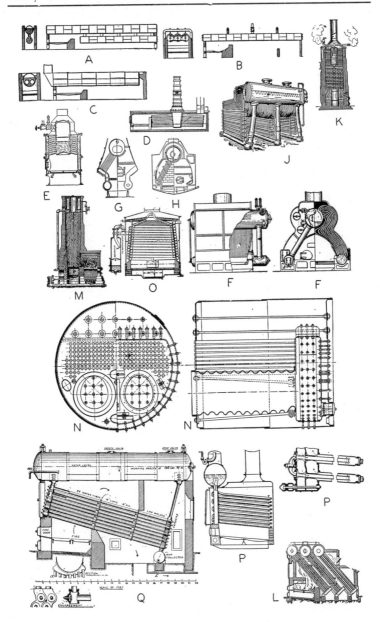

CLASS VII.   COMBUSTION

Section 44c.   Early Stokers and Furnace Grates

A—Tupper furnace grate with dumping sections.

B—Shaking grate for a boiler furnace.

C—Shaking and tipping furnace grate.

D—Jones underfeed stoker.

E—Playford mechanical stoker.

F—Meissner mechanical stoker.

G—Rocking-grate stoker (English model); typical American models of the same period were the "Roney" and the "Taylor."

H—Green chain-grate stoker setting arrangement under a horizontal-return tubular boiler.

J—Hopper feed (Leach Feuerung, Germany).

K—Chain grate (Steinmuller-Kettenrest, Germany).

L—American stoker; spiral-screw feed.

M—Coking-type stoker; the hoppers contain small coal, placed there by a mechanical elevator; the coal is fed slowly forward by a screw then coked at C, and pushed forward gradually by the movement of the fire bars, which are connected to the crank shaft; finally the fuel is completely burnt, and drops down as ash (New Conveyor Co., Great Britain).

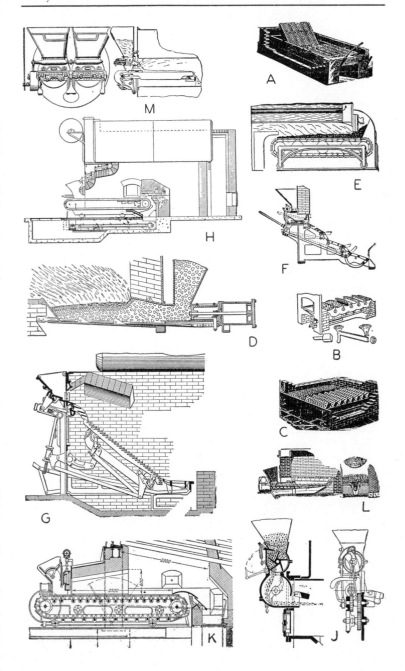

CLASS VII.   COMBUSTION

Section 44d.   Modern Mechanical Stokers

A—Hopper-type underfeed stoker; used primarily for home heating.

B—Bin-type underfeed stoker; used for home heating.

C—Bin-type underfeed anthracite stoker with automatic ash removal.

D—Stoker-fired air-conditioning unit.

E—Hopper-type underfeed screw stoker.

F—Bin-type underfeed screw stoker.

G—Underfeed side-cleaning stoker.

H—Overfeed traveling-grate stoker.

J—Pneumatic overfeed spreader stoker.

K—Rotor-type overfeed stoker.

L—Fire-tube boiler (Skelly Stoker and Combustion Engineering Co.).

M—Chain-grate stoker with Stirling boiler, superheater, economizer and air heater over a water-cooled furnace (Babcock and Wilcox).

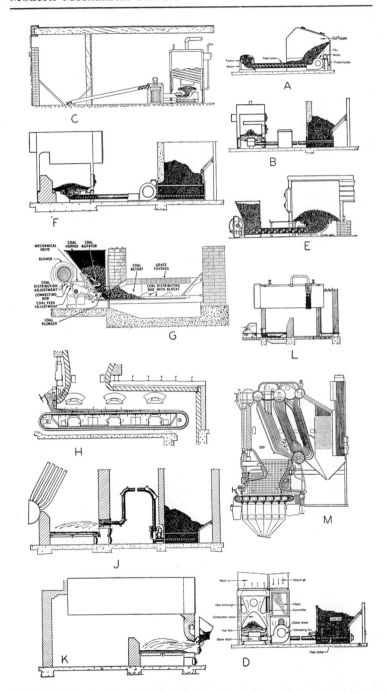

CLASS VII.   COMBUSTION

Section 45a.   Vaporizing Oil Burners

A—Pot-type vaporizing burner.

B—Range burner installation.

C—Fuel-oil strainer.

D—Typical gas generator vaporizing burner.

E—Natural-draft vaporizing burner.

F—Natural-draft vaporizing burner installed.

G—Vaporizing on hot-spot plate.

H—Vaporizing by conducted heat.

J—Portable oil-burning torch.

K—Automatic electric preheater for fuel oil

L—Radiant-heat oil burner.

M—Multitube, multipass preheater.

N—Multicoil preheater.

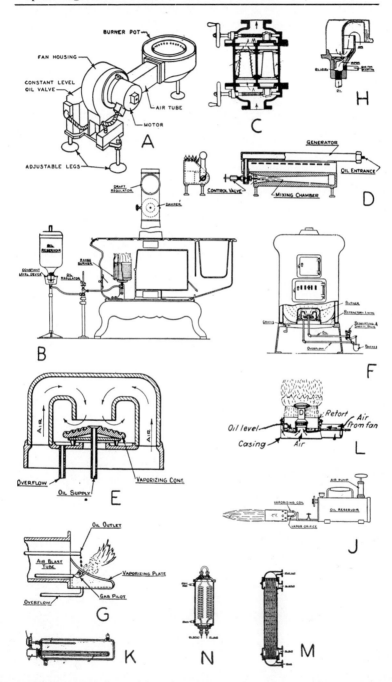

CLASS VIII.   PRIME MOVERS

Section 46a.   Nineteenth-Century Engines

A—Newcomen's engine first built at Wolverton, England for pumping water from a mine in the year 1711; James Watt (1736-1819), however, introduced so many new features that he is usually regarded as the inventor of the modern reciprocating steam engine; the three valves were manipulated by hand: one to admit steam to raise the piston; the second to admit a spray of cold water which condensed the steam, causing the atmospheric pressure to push the piston down; the third was open to exhaust; the cycle was then repeated; later a valve motion was developed for continuous operation.

B—Steeple engine with guide and crosshead; very early type.

C—Trunk engine; it eliminates the crosshead and guide; very early type.

D—Diagonal twin-screw engine with cross-over connecting rods economizing space in vessels.

E—Twin-screw vertical-cylinder engine; the inner gears keep the beam even; the outer gears are on propeller shafts; early type.

F—Inclined paddle-wheel engine with upright crank-connected beam for operating a pump.

G—Vertical engine with bell-crank lever for propelling a stern-wheel boat.

H—Oscillating engine with trunnions on the middle of the cylinder.

J—Twin-screw oscillating engine with through piston rod; early type.

K—Compound oscillating engine.

L—"Brotherhood" three-cylinder engine; steam, admitted to the central chamber, develops equal pressure on all pistons; a rotary-disc valve operated by a crank pin admits steam to the outside of the pistons.

M—Tandem compound vertical engine with two piston rods for the low-pressure cylinder (1904).

N—Tandem compound vertical engine (1904).

O—Compound engine for twin-screw propellers.

P—High-speed tandem compound engine; "Harrisburg" model (1920).

Q—The Belliss high-speed engine for direct connection to an electric generator (1910).

R—Willans central-valve engine (1910).

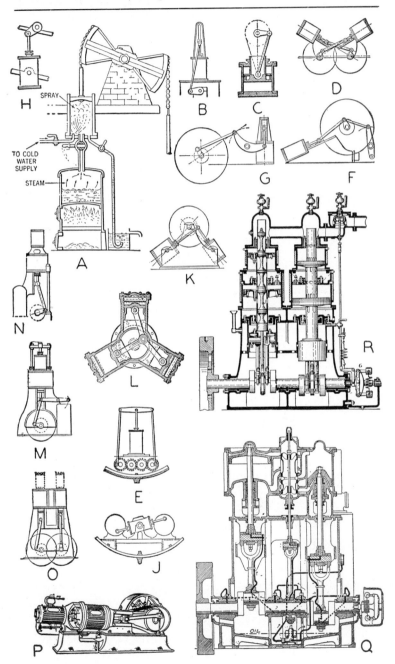

CLASS VIII. PRIME MOVERS

Section 46b.   Steam-Engine Valves and Valve Gears

A—Simple single D slide valve without lap.

B—Valve with steam lap and exhaust lap.

C—Single D slide valve with double exhaust and steam ports; central steam ports open into the steam chest at the side of the valve.

D—Double-ported slide valve.

E—Gridiron slide valve for large port area with minimum motion of the valve.

F—Balanced slide valve; a ring in a recess of the valve rides against the steam chest cover, held in place by a spring.

G—Balanced slide valve, Buchanan and Richter's patent.

H—Balanced slide valve.

J—Steam-engine valve chest (Erie City Iron Works).

K—Variable-expansion valve gear (Meyer).

L—Double-ported slide valve, with relief ring.

M—Back cutoff valve with double ports (Main).

N—Straight-line valve.

O—Double ports to both Meyer and Main valves.

P—Modern high-speed engine valve.

Q—Balanced throttle valve with direct governor connection.

R—Multiple-ported piston throttle valve.

S—Butterfly or wing throttle valve directly connected to the governor.

T—Piston-balanced valve.

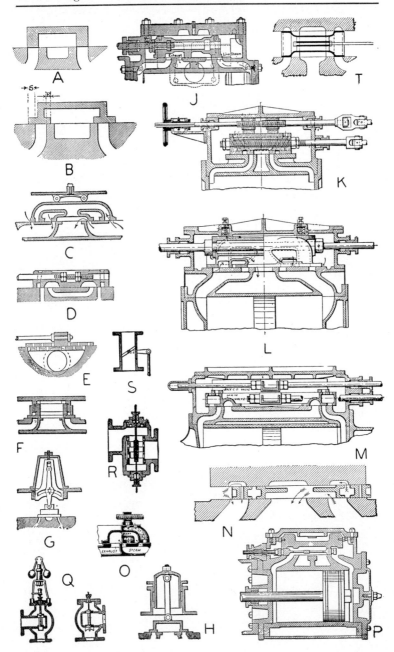

CLASS VIII.　PRIME MOVERS

Section 46c.　Link-Motion and Valve Gears

A, B—Stephenson's link motions.

C—Gooch's link motion.

D—Allan's link motion.

E—Joy's radial-valve gear.

F—Variation of Joy's valve gear.

G—Walschaert's valve gear.

H—Hackworth's valve gear.

J—Marshall's valve gear.

K—Locomotive link motion.

L—Reversing link motion.

M—Valve gear of an oscillating marine engine.

N—Single-eccentric valve gear with variable travel, adjusted by a hand wheel; the eccentric drives a block in a slotted link, which is rocked on a central pivot by the screw to change the throw of the valve.

O—Cam-bar valve movement.

P—Bremme valve gear.

Q—Valve gear of a Cornish engine with trip puppet valves.

R—Fink link gear for a D slide valve; single-eccentric variable valve throw.

S—Variable expansion valve gear with one eccentric.

T—Tappet-lever valve motion; used on pumps and rock drills.

U—Valve motion eccentric.

V—Shifting eccentric.

W—Misch's valve tappet for a steam pump; a three-armed lever rocked by a roller traveling with the piston rod.

X—Corliss valve gear and governor.

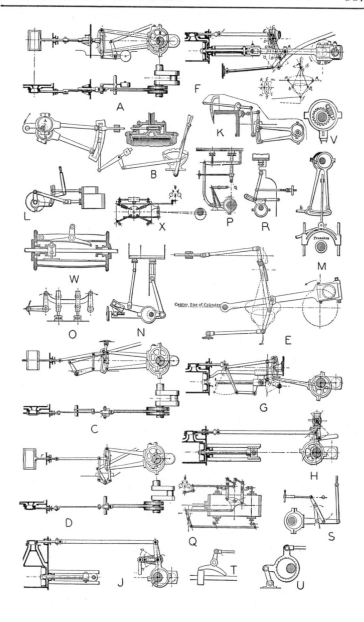

CLASS VIII.   PRIME MOVERS

Section 47a.   Early Rotary Engines

A—Cochrane rotary engine; a wing piston is rotating around the central axis of an outer cylinder or shell; a hollow cylinder of smaller diameter is pivoted eccentrically to the wing axis to keep one side in contact with the shell; the steam pressure revolves the wing and shaft with a force due to the varying area of the wing outside the inner cylinder.

B—Another model of the Cochrane engine.

C—Franchot rotary engine.

D—Double-slide-piston rotary engine.

E—Simple rotary engine.

F—Lamb rotary engine.

G—Napier rotary engine.

H—Roller-piston rotary engine.

J—Boardman rotary engine.

K—Variation of the Cochrane engine.

L—Rotary engine with wing barrel and concentric shaft.

M—Berrenberg rotary engine.

N—Smith rotary engine.

O—Fletcher's rotary engine.

P—Bartrum and Powell rotary engine.

Q—Stocker rotary engine.

R—Forrester rotary engine.

S—Kipp rotary engine.

T—Ruth's rotary engine.

U—Almond rotary engine.

V—Rotating-cylinder engine.

W—Rotary multicylinder engine.

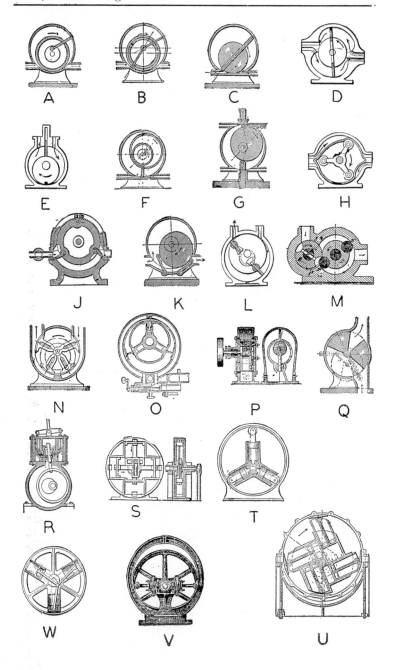

A   B   C   D

E   F   G   H

J   K   L   M

N   O   P   Q

R   S   T

W   V   U

CLASS VIII.   PRIME MOVERS

Section 47b.   Early Rotary Engines

A—Davies disc rotary engine.

B—Link vibratory rotary engine.

C—Revleaux rotary engine or pump.

D—Knickerbocker four-piston rotary engine.

E—Root's double-quadrant engine.

F—Root's square-piston engine.

G—Wilkinson's steam turbine.

H—Dow steam turbine.

J—De Laval steam turbine; a jet or jets of steam impinge at a small angle on the concave buckets at the periphery of a disc wheel, pass through and exhaust at the other side; the nozzle has an expanding orifice.

K—Plan of the De Laval turbine, showing the nozzle.

L—Parsons steam turbine; a series of discs is fixed on a revolving shaft; intersecting discs are fixed on the stationery shell; the face of the revolving discs has small blades set at an angle to the radius; the stationary discs have a similar set of blades interlocking with the revolving blades and set at a contrary angle; steam passes from the valve to the inner edge of the first disc, then outward through the blades; then it returns through the vacant space of the next pair and streams outward again.

M—Rotary engine with eccentric piston and two sliding vanes or steam stops.

N—Rotary engine; an eccentric ring revolves on its center, carrying around an eccentric boss and vanes, allowing the vanes to alternately project into the steam space as the wheel revolves.

O—Ivory's rotary engine having an eccentric cam and two sliding shutters, with a central steam inlet.

P—Bisschop's single-acting disc engine with three or four cylinders, whose rams press alternately on the edge of a disc.

Q—Another type of disc engine in which partitions, rising and falling vertically, form the steam stops.

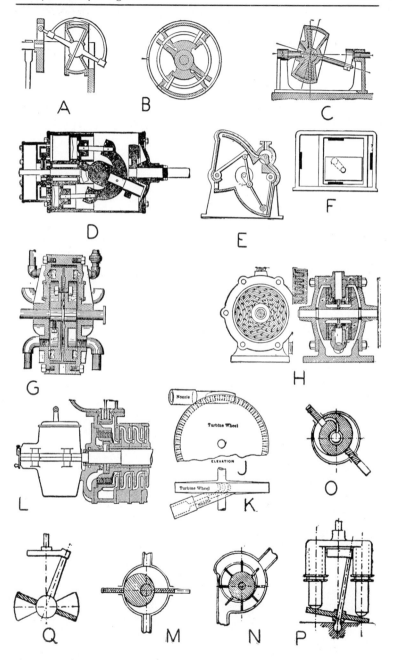

A

B

C

D

E

F

G

H

Nozzle

Turbine Wheel

O

ELEVATION J

Turbine Wheel

Nozzle

K

L

Q

M

N

P

CLASS VIII.   PRIME MOVERS

Section 48a.   Modern Steam Turbines

A—The high-pressure element of a cross-compound main-propulsion impulse turbine; a sight-flow indicator and a thermometer are provided in the oil return from each main bearing, which are of the self-aligning, babbitted type; the turbine shaft, at the steam-inlet end, is sealed by combination labyrinth and carbon-ring packing, while at the exhaust end, carbon-ring packing seals are provided.

B—Low-pressure element of a cross-compound main-propulsion impulse turbine and impulse-astern turbine; the astern turbine has two impulse stages, the first of which is velocity compounded; sight-flow indicators and thermometers are provided in the oil discharge from each bearing.

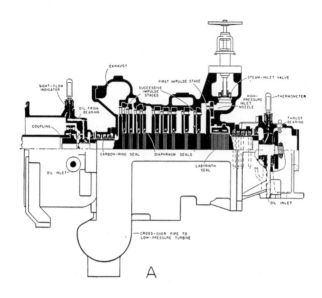

A

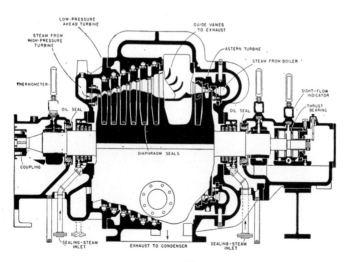

B

CLASS VIII.   PRIME MOVERS

Section 49a.*   **Principles of Gas Turbines**

A—Heron's gas turbine from 130 B.C.

B—The "smoke jack" from the 17th century.

C—The steam engine.

D—Steam turbine.

E—Aviation-gasoline or diesel engine.

F—Gas turbine.

G, H—Relation between a typical domestic oil burner and the elementary gas turbine; both the blower and generator are driven by the gas turbine.

J—Single turbine unit without heat exchanger; rows of blades deliver a considerable amount of useful power to the propeller; rows of stationary blades alternate with rows of rotating blades; each stationary blade redirects the hot gases or air against the passing rotating blades; this arrangement is known as the simple cycle.

K—Single turbine unit with heat exchanger; known as the regenerative cycle.

L—Engine with exhaust-gas-driven turbo-charger; the arrows show the movement of the air into the combustion space, forcing the exhaust gases out and into the turbine.

*The subject matter and illustrations were taken by permission from the Modern Gas Turbine by R. Tom Sawyer, Prentice-Hall Inc.

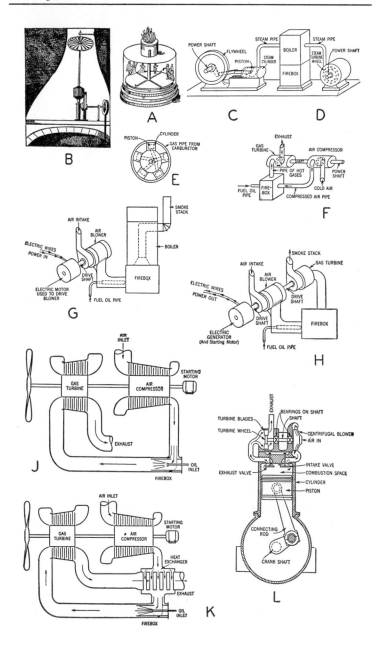

CLASS VIII.   PRIME MOVERS

Section 49b.*   Jet Propulsion

A–Campini design of high-altitude craft for operation at either subsonic or supersonic speeds. The control cabin is pressure charged; *A* ovoid cabin; *B* enshrouding cylinder; *C* two-stage centrifugal compressor; *D* radial engine; *E* rectifier-radiator; *F* combustion space; *G* annular mixing channel; *H* discharge nozzle; *J* cone for varying the nozzle orifice; *K* controlled lateral orifices; *L* slidable shroud ring.

B–Gas turbine-jet design in Air Commodore Frank Whittle's first jet-propulsion patent; *1* air in; *2* rotating portion of the compressor; *3* combustion chamber, showing the oil spray; *4* turbine wheel; *5* exhaust jet.

C–-Jet-propulsion design made by A. Lysholm and sponsored by Milo Aktiebolaget, of Stockholm, Sweden; the streamlined Milo unit is intended for wing installation; all the air compressed by the multistage blower is passed to the combustion chamber and, with added fuel, is expanded through the multistage gas turbine; *A* four-stage centrifugal compressor; *B* air chamber; *C* combustion chamber; *D* turbine; *E* fuel injector; *F* discharge nozzle.

D–Sweden's Ljungstrom's gas turbine unit embodying dual twin-rotor blowers for compressing the air; special arrangements are made for speeding-up the compressor to permit rapid acceleration of the aircraft at the take-off or in emergency; *A* screw-type, twin-rotor compressors; *B* air chamber; *C* combustion chamber; *D* turbine; *E* discharge conduit; *F* upper auxiliary conduit; *G* lower auxiliary conduit.

E–Self-contained all-rotary jet-propulsion unit patented by Max Hahn and assigned to the firm of Ernst Heinkel, Germany; *A* blower impeller; *B* turbine wheel; *C* air-flow guide ring; *D* annular combustion chamber; *E* combustion-air passage; *F* fuel injector; *G* insulation passage.

F–Ljungstrom's gas-turbine screw and jet-propulsion unit with multistage centrifugal compressor; *A* centrifugal compressor; *B* air inlets; *C* annular air duct; *D* combustion chamber; *E* bulkhead; *F* fuel nozzles; *G* gas turbine; *H* discharge duct; *J* discharge flap; *K* effluent by-pass flap.

*Reprinted by permission of Aerosphere, Inc., from *Gas Turbines and Jet Propulsion for Aircraft* by G. Geoffrey Smith.

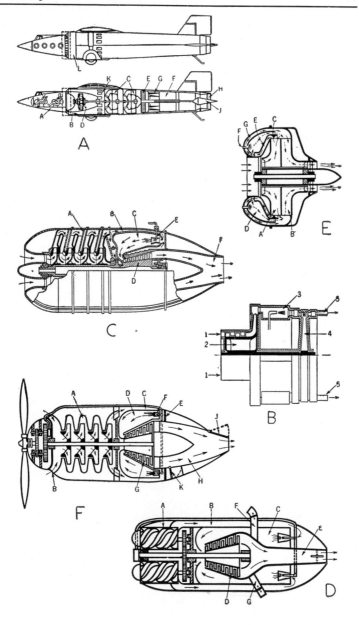

CLASS VIII. PRIME MOVERS

Section 49c. Rockets and Jet Propulsion

A—Dry-fuel reaction-rocket motor; true rocket type.

B—Liquid-fuel reaction-rocket motor; true rocket type.

C—Thermal-jet engine (Turbojet); air-stream engine type.

D—Intermittent duct engine (buzz-bomb engine); air-stream engine type.

E—Continuous duct engine (athodyd); air-stream engine type.

*F—Simple sky rocket.

G, H—Sounding rocket with gyro control having regenerative motor and collapsible fuel bag; for vertical flight carrying instruments; a prewar design, now obsolete.

J—Rocket trajectory of G, H.

K—Jet-propulsion action of solid-propellant rocket motor.

L—Jet-propulsion action of liquid rocket system.

*Figures F to L are reprinted by permission from a paper by Lovell Lawrence.

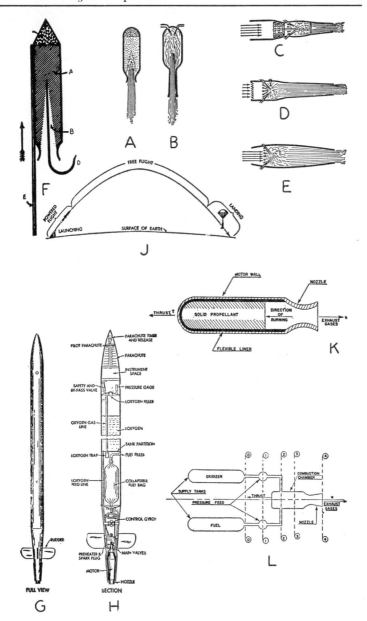

CLASS VIII.   PRIME MOVERS

Section 50a.   Diesel-Engine Principles

A—Four-cycle suction stroke; the air is drawn into the cylinder through an open inlet valve by the retreating piston; *a* inlet valve; *b* fuel nozzle; *c* exhaust valve.

B—Four-cycle compression stroke; after the piston has passed the dead center, the air-inlet valve closes, the piston starts to return, and compression begins; shortly before the end of this stroke, the fuel valve opens and injection commences; meanwhile, the compression of the trapped air has raised its temperature as high as 1000 to 1200°F, so that ignition of the fuel takes place as soon as it is injected either by mechanical or air injection.

C—Four-cycle power stroke; combustion continues during the early part of the succeeding stroke until after the closing of the fuel valve; the temperature reaches about 3000°F (1649°C); after the fuel valve closes and combustion ceases, the gases expand during the remainder of the stroke; the exhaust valve opens somewhat before dead center.

D—Four-cycle exhaust stroke; burned expanded gases are expelled during the fourth piston stroke; just before dead center has been reached, the air-inlet valve opens, and the cycle is repeated.

E—Two-cycle compression stroke; at the beginning of this stroke, the exhaust ports are still uncovered; at the same time, scavenging air is entering the cylinder; the scavenging air ports are closed by the advancing piston; the exhaust ports are likewise closed an instant later; compression follows during the remainder of the stroke; just before dead center, the fuel valve opens and injection begins.

F—Two-cycle power stroke; combustion stops considerably later than when the piston has passed dead center, and the burned gases expand during the remainder of the power stroke; when the exhaust ports are again uncovered by the retreating piston, the gases are released to the atmosphere; an instant later, the cylinder is purged either by the uncovering of the scavenging air ports or by the opening of scavenging air valves; the cycle is then repeated during the next crank revolution; *a* scavenging-air port; *b* fuel nozzle; *c* exhaust port.

G—Engine scavenging with opposed piston.

H—Scavenging through exhaust valves in head.

J—Junker's four-cylinder cycle events.

K—Mechanical-injection fuel system.

L—Air-injection fuel system.

M—Rotary-type scavenging air blower with end plate removed.

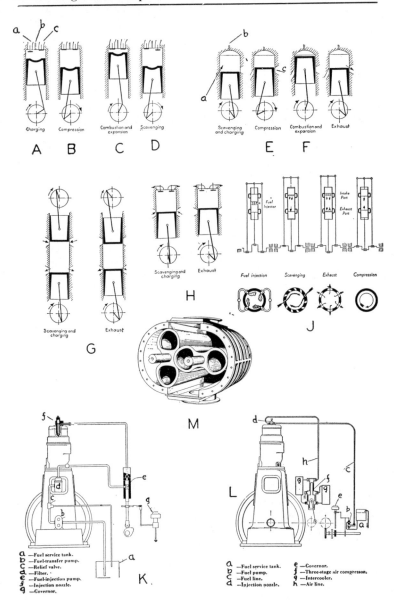

A Charging   B Compression   C Combustion and expansion   D Scavenging

E Scavenging and charging   F Compression   Combustion and expansion   Exhaust

G Scavenging and charging   Exhaust

H Scavenging and charging   Exhaust

Fuel Injector   Intake Port   Exhaust Port

J Fuel injection   Scavenging   Exhaust   Compression

M

K
a —Fuel service tank.
b —Fuel-transfer pump.
c —Relief valve.
d —Filter.
e —Fuel-injection pump.
f —Injection nozzle.
g —Governor.

L
a —Fuel service tank.   e —Governor.
b —Fuel pump.   f —Three-stage air compressor.
c —Fuel line.   g —Intercooler.
d —Injection nozzle.   h —Air line.

CLASS VIII.   PRIME MOVERS

Section 51a.   Elementary Valve Gears for Gas Engines

A—Four-cycle gas engine with half-reducing gear; push-rod lever and two push rods for regulating charging and exhausting.

B—Valve gear for gas engines for opening the exhaust valve of a four-cycle engine; the eccentric *A* gives the push rod *D* a forward stroke at each revolution of the shaft; the ratchet wheel *C* has a friction resistance, every other tooth being a shallow notch so as to hold up the lip of the push rod at every second revolution of the shaft and make a miss-hit on the valve rod; at the next revolution, the lip falls into the deep notch and the push rod opens the exhaust valve.

C—Valve gear for a four-cycle gas engine; the two-thread worm *a* on the engine shaft has the middle part of the thread extended to form a cam; the gear *B,* revolves and at alternate revolutions, the cam section of the worm runs in a recess of the revolving gear, and the valve rod is not operated, thus opening the exhaust valve at every second revolution.

D—Plumb-bob gas-engine governor; the plumb-bob *A* is pivoted in a box attached to the exhaust valve push rod; back motion of the push rod causes forward motion of the bob acting as a pendulum, and a downward motion of the pick blade *C*, bringing it in contact with the valve spindle *D;* the screws *E* and *F* are for adjustment of the motion of *A.*

E—Double-grooved eccentric for two lengths of rod thrown alternately by traversing the push rod in the cross grooves; used also for single-valve rod throw of four-cycle engines.

F—Valve gear for four-cycle gas engines; the cam is keyed to the engine shaft; the inner ring gear is swept around within the outer fixed gear, skipping one tooth at each shaft revolution; this makes a contact of a ring-gear tooth with the exhaust-valve rod at every other revolution, necessary for the operation of a four-cycle engine.

G—Inlet valve for a gas engine.

H—Inertia governor for hit-miss governing.

J—Differential cam throw; the rolling disc is traversed by the governor from one cam to another.

K—Gas-engine valve gear; *E* inlet valve; *F* exhaust valve; they are operated by a bent lever, with sliding roller *H* and double cam *C,* which, by the groove, rides the roller alternately on, to the cams.

L—Governor and variable cam for varying the movement of an inlet valve by fly-ball action.

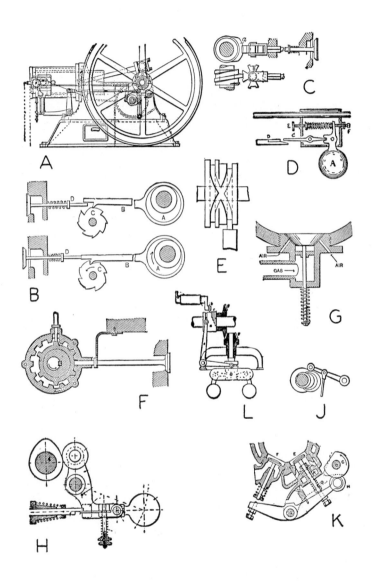

CLASS VIII.   PRIME MOVERS

Section 52a.   Principles of Hydraulic Power

A, B—Gravity machines; they are rarely used but illustrate the principle.

C—Overshot wheel; a gravity machine with steel buckets.

D—"Leffe" overshot wheel.

E—Overshot wheel.

F—A more efficient overshot design.

G—Oscillating-cylinder pressure engine.

H—Pressure-type hydraulic riveter.

J—Hydraulic cylinder with Thorp controller for hydraulic lifts.

K—Pelton wheel; typical velocity machine.

L—Double bucket for improving efficiency of a Pelton wheel due to a cut-out lip.

M—Buckets of a Pelton wheel, showing the method of separating jet and returning the parts nearly in line with the impact jet, thus gaining about 85 per cent of power.

N—Breast water wheel; the power is about 30 per cent of the fall.

O—Undershot water wheel with a power about 30 per cent of water fall; used with heads less than 6 feet.

P—Saw-mill water wheel.

Q—Flutter water wheel; it has very low efficiency.

R—Current wheel; it has low efficiency.

S—Barker wheel for small power; low-efficiency reaction type.

T—Breast wheel.

U—Water-wheel governor.

V—Pelton-wheel buckets.

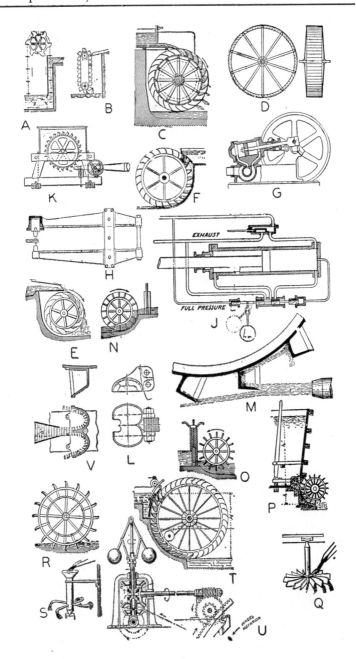

CLASS VIII. PRIME MOVERS

Section 52b. Hydraulic-Turbine Principles

A—Jonval turbine; *a* chute, guide vane or fixed directrix; *c* bucket or runner.

B—Jonval turbine; downward flow; either the upper or the lower set of vanes may be fixed.

C—Lancaster turbine; downward discharge; the upper parts of the blades are vertical and receive water tangentially from the gate plates.

D—Model turbine; the runner has downward discharge; register gates are pivoted and operated by arms from a sector.

E—Swain turbine; inward and downward flow, with inward-curved vanes or flumes.

F—Camden turbine; two independent sets of buckets; the upper set is inward and central discharge; the lower set is curved backward, with tangential discharge.

G—Turbine and gate; downward flow from angular fixed guide vanes in the water chamber.

H—Munson double turbine; the water discharges both upward and downward through curved guide vanes to reverse curves in top and bottom runner wheel vanes.

J—Warren central-discharge turbine; the wheel revolves inside the fixed directrix; water enters from outside and discharges into and beneath the wheel.

K—Fourneyron turbine; the rim of the outer buckets is revolving around the inner fixed directrix, the water flowing outward.

L—Leffel double-runner turbine; the upper section of the runner discharges inward and down the center; the lower section has curved blades to discharge downward; there is one register gate for both sections.

M—Properly formed guide vanes for impulse wheels.

N—Improperly formed guide vanes for impulse wheels.

O—Apparatus for measuring the force of a water jet; the total force measured by the weight *w; v* is the direction of force.

P—Right-angle jet; the vertical force is measured by a platform scale, the horizontal force, by a spring scale.

Q—Governor.

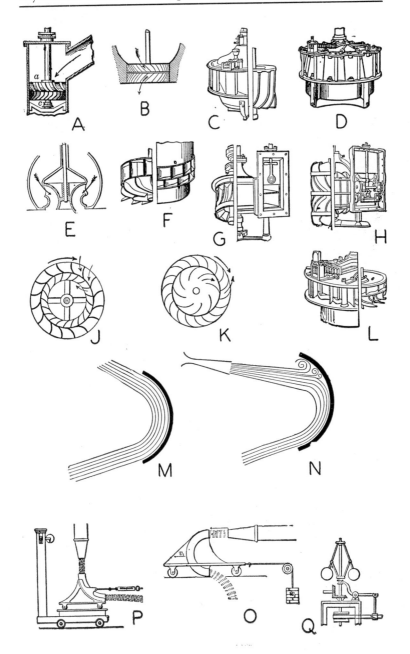

CLASS VIII.   PRIME MOVERS

Section 53a.   Windmills

A—Early Dutch type windmill with reefing sails.

B—Windmill and steel tower; the tail piece is swung around to turn the face of the windmill from the wind by a governor.

C—Modern windmill.

D—Hemispherical cup mill; used as anemometer.

E—Early-type windmill on a two-stone flour mill.

F—Windmill operating an electric generator.

G—Curved-vane windmill.

H—Feathering windmill.

J—Box kite.

K—Kansas jumbo mill; it can irrigate 6 acres with a simple pump connection.

L—Spiral-vane coil for a ventilator top.

M—Windmill with adjustable sails.

N—Spiral wind wheel.

O—Nance's windmotor; a pair of endless chains carries pivoted curved vanes which are so arranged as to deflect the wind from one series to another.

P—Pantanemone; two sails at right angles, as shown, will drive a diagonal shaft with the wind from any direction.

Q—Governor of a propeller windmill (Wincharger Corporation).

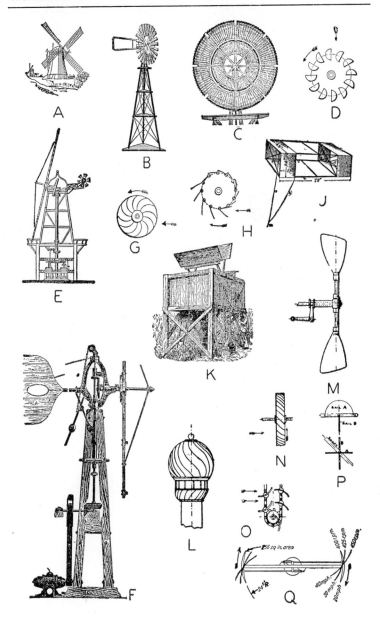

A

B

C

D

E

G

H

J

F

K

L

M

N

O

P

Q

CLASS VIII.   PRIME MOVERS

Section 53b.   Compressed-Air Power

A—Proper layout of air-compressor plant for operating pneumatic tools; the air must be clean and dry (Ingersoll-Rand Co.).

B—Clayton compressed-air lift.

C, D—Pneumatic hammers.

E—Pneumatic riveter.

F—Pneumatic hammer (Hotchkiss).

G—Pneumatic hammer (Grimshaw).

H—Portable riveter.

J—Pneumatic breast drill.

K—Oscillating-cylinder pneumatic drill.

L—Pneumatic drill with rotary motor.

M—Pneumatic sheep-shearing machine.

N—Pneumatic caulking tool (Great Britain).

O—Pneumatic hammer in position to strike (Great Britain).

P—Caulking chisel.

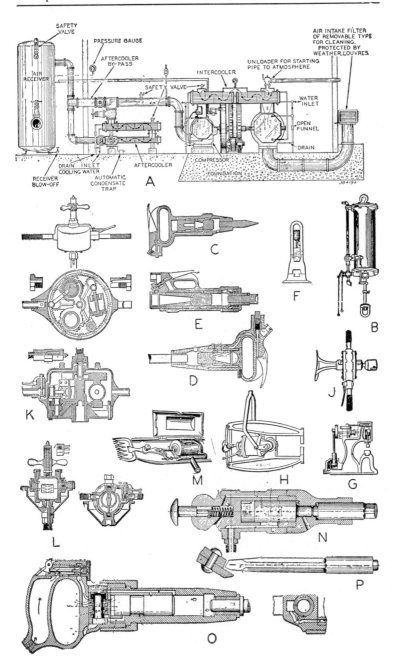

SAFETY VALVE

PRESSURE GAUGE

AFTERCOOLER BY-PASS

AIR RECEIVER

AIR INTAKE FILTER OF REMOVABLE TYPE FOR CLEANING. PROTECTED BY WEATHER LOUVRES.

UNLOADER FOR STARTING PIPE TO ATMOSPHERE.

INTERCOOLER

SAFETY VALVE

WATER INLET

OPEN FUNNEL

DRAIN

DRAIN INLET COOLING WATER

RECEIVER BLOW-OFF

AUTOMATIC CONDENSATE TRAP

AFTERCOOLER

COMPRESSOR

FOUNDATION

A

C

F

B

E

D

J

K

M

H

G

L

N

P

O

CLASS VIII. PRIME MOVERS

Section 53c.   Sea-Wave Power

A to G—Various devices to derive power from wave motion
    by rise and fall of a float.

H to M—Swinging blades anchored in various ways. An-
    chored floats and the motion of two boats may be
    used for a variety of applications.

N, O—Swinging motion of the waves near the shore trans-
    mitted by means of large blades swung from a pier:
    N shows single-acting transmission; O shows dou-
    ble-acting transmission.

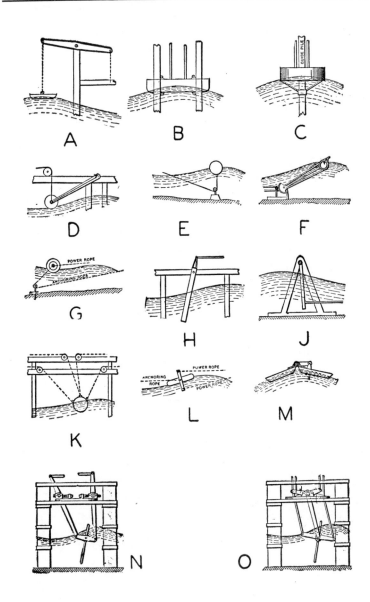

A

B

C

D

E

F

G

H

J

K

L

M

N

O

CLASS IX. TRANSPORTATION

Section 54a. Wheels

A—Huxley's wheel with spring tire and jointed spokes

B—Double tires and intermediate spring.

C, D—Bent spoke, spring formation.

E—Outer elastic tire and inner rigid ring to which the springs are attached.

F—Part of an elastic wheel consisting of steel spokes and a transverse coil of steel forming the ring.

G—Part of an elastic wheel; the rim contains radial air cylinders; the plungers are fixed to the tire and kept open by springs.

H—Rubber insert between the rigid inner ring and the outer elastic ring.

J—Section of solid rubber tire.

K—Spring wheel with solid or pneumatic tire.

L—Section of a wire wheel spoke and rim for pneumatic tire.

M—Free-running axles for mining cars; the divided axle is held together by grooved bearings.

N—Street-car truck with spring frame and brake connections.

O—Street-car truck; the larger wheel is geared to the motor while the small wheel is the trailer.

P—Old-style freight-car truck showing the forward half with the brake beam, safety chain, spring, and bearing bar.

Q—Landing wheels of a fighter plane; the wheels are retractable and equipped with pneumatic tires, shock absorbers, and hydraulic brakes.

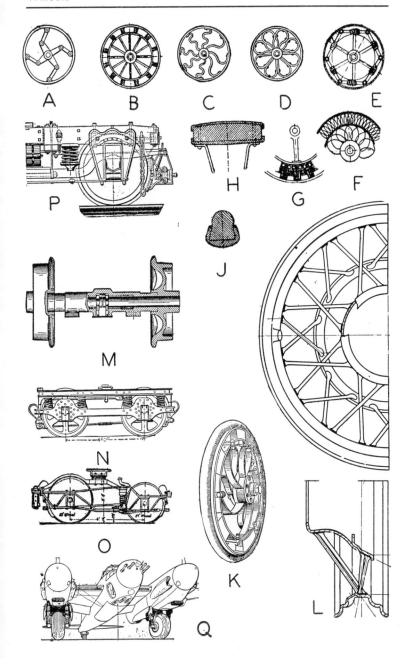

A B C D E

P H G F

J

M

N

O

K

L

Q

CLASS IX.   TRANSPORTATION

Section 54b.   Vehicle Bodies

A—Furniture van.

B—Inclined passenger car; used on very steep hills.

C—Hopper wagon with central discharge.

D—Long truck for boilers, tanks, etc.

E—Tip wagon with three centers.

F—Passenger arrangement; transverse seats and central aisle.

G—Passenger arrangement; outside and central longitudinal seats.

H—Passenger arrangement; upper and lower longitudinal seats.

J—Passenger arrangement; reverse of that in H.

K—Passenger arrangement for monorail.

L—Passenger arrangement; reverse of that in G.

M—Passenger arrangement open; with either transverse or longitudinal seats.

N—Open freight car or wagon.

O—Covered freight car or wagon.

P—Hopper-car or wagon; discharging below.

Q—Side-discharge hopper wagon.

R—Side-tip three-center wagon.

S—Rocker-side dump car.

T—Gable-bottom car.

U—Scoop dump car.

V—Hopper-bottom car.

W—Box-body dump car.

X—Cross-over dump; car unloading.

Y—Rotary gravity dump.

Z, AA, BB—Differential air dump car three views shown; double trunnion, double fulcrum; dumps either side (Differential Steel Car Co.).

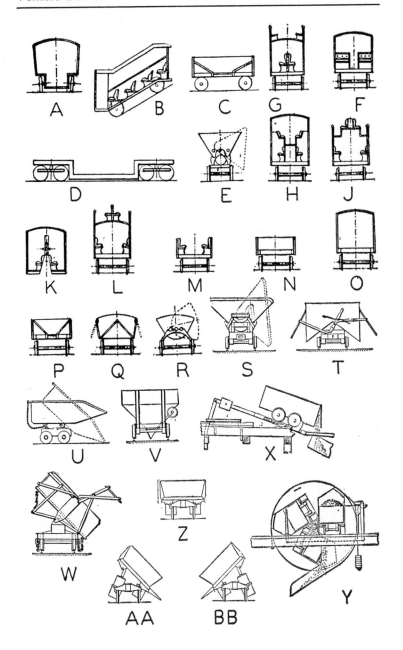

A B C G F
D E H J
K L M N O
P Q R S T
U V X
W Z Y
AA BB

CLASS IX.   TRANSPORTATION

Section 54c.   Underframes and Trucks

A—Two-wheel suspension car for single rail or wire rope.

B, C, D, E—Three-wheel cars; tricycles.

F, G, H—Different forms of four-wheel underframes, with and without swiveling bogies.

J—Four-wheel car, with the leading and trailing wheels off the ground, it is used as a hand truck, and is easily swiveled about, running actually on three wheels only.

K—Five-wheel underframe, with and without swiveling trucks.

L, M—Plans of six-wheel cars, with swiveling gear for curves; the center pair has end play and swivels the leading and trailing axles by means of the jointed braces or stays.

N—Plan of a four-wheel car, with swiveling gear for curves.

O, P, Q—Six-wheel cars, with leading and trailing swiveling trucks. In P and Q, the center pair of wheels, if running on rails, must have broad, flat tires.

R—Eight-wheel double-truck underframe, usually employed in long cars; each truck is free to swivel independently and is centrally loaded.

S—Ten-wheel double-truck underframe; the center pair must have end play or broad flat tires.

T—Twelve wheels and three trucks; the center truck must have end play as in P or S, or transverse rollers between the truck and frame.

U—Segmental swiveling bearings, used instead of a swiveling truck and center pin.

V—Swiveling gear for car wheels.

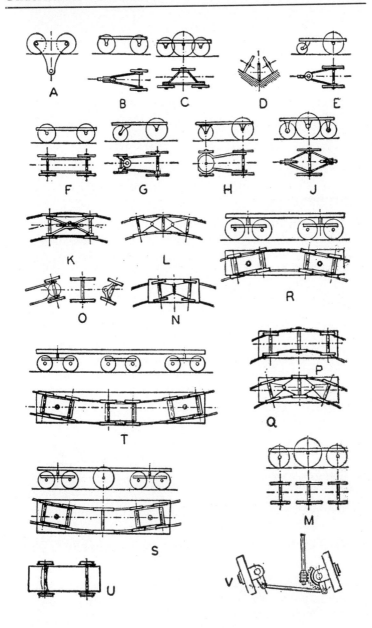

CLASS IX.   TRANSPORTATION

Section 55a.   Sailboat Types

A—Skipjack or catboat with a baggy mainsail 5 bent to the mast.

B—Leg-of-mutton sail with a triangular mainsail 5 attached to the mast and boom.

C—Lateen rig with a triangular mainsail 5 extended by a long yard, which is slung about one-quarter of its length from the lower end.

D—Square or lug sail 5 attached to the yard.

E—Split-lug or square sail attached to a yard and divided at the mast, the lower portion being bent to the mast; it has a mainsail 5 and a jib 2.

F—Newport-type catboat; similar to A but with a forestay to a short bowsprit; it has a mainsail 5.

G—Sloop; it has a mainsail 5 and a jib 2 with fore- and back-stays.

H—Two-masted or dipping lug with a foresail 9 and mainsail 5 which are square, except at the top where they are bent to the yards hanging at a slant.

J—Lateen rigged felucca; two-masted boat with a lateen foresail 4, a mainsail 5 and a jib 2.

K—Pirogue; two-masted schooner rig without jib and a leeboard or centerboard; it has a foresail 4 and a mainsail 5.

L—Three-quarter lug rig; it has a jib 2, a foresail 4, and a mainsail 5.

M—Jib-topsail sloop; it has a flying jib, a jib 1, another jib 2, a mainsail 5, and a gaff topsail 13.

N—Schooner rig with a jib 2, foresail 4, and mainsail 5.

O—Skiff yawl rig having a flying jib 1, a jib 2, a mainsail 5, and a lugsail 8.

P—Sliding gunter with a sliding topmast, a jib 2, a foresail 4, and a mainsail 5.

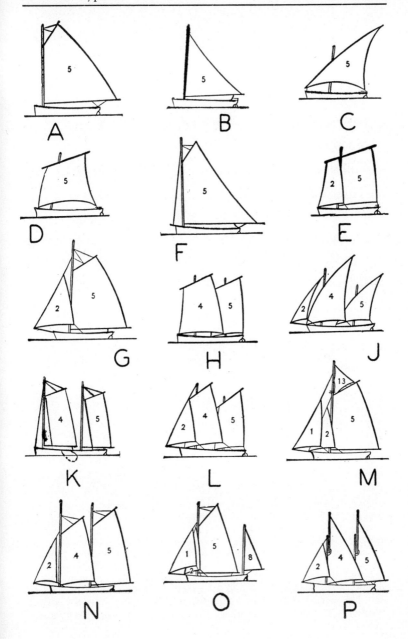

CLASS IX.   TRANSPORTATION

Section 55b.   Sailboat Types

A—Cutter with sails *1, 2, 5,* and *9.**

B—Sloop yawl with sails *2, 5* and *8.*

C—Full schooner rig with sails *1, 2, 3, 4, 5, 12* and *13.*

D—Topsail schooner with sails *1, 2, 4, 5, 9* and *13.*

E—Club topsail rig with sails *1, 2, 4, 5, 12, 13, 14* and *31.*

F—Hermaphrodite brig with sails *1, 2, 4, 5, 9, 13, 14, 22, 25* and *32.*

G—Brigantine with sails *1, 2, 4, 5, 9, 10* and *22.*

H—Barkentine with sails *1, 2, 3, 4, 5, 7, 9, 13, 14, 22, 25, 32, 33* and *34.*

J—Full-rigged brig with sails *1, 2, 3, 4, 5, 7, 9, 10, 20, 22, 25* and *32.*

K—Bark with sails *1, 2, 3, 4, 5, 7, 16, 17, 19, 20, 22, 23, 25, 26* and *34.*

L—Full-rigged ship with sails *1* to *30* inclusive, and *35.*

M—Full-rigged ship with sails *1, 2, 3, 4, 5, 6, 9, 10, 11, 22, 23, 24, 25, 26* and *27.*

*The sail types are represented by the following numerals; *1* denotes a flying jib; *2,* jib; *3,* foretop staysail; *4,* foresail; *5,* mainsail; *6,* cross jacksail; *7,* spanker; *8,* lugsail; *9,* topsail or foretopsail; *10,* main topsail; *11,* mizzen-topsail; *12,* foregaff-topsail; *13,* gaff topsail or main gaff topsail; *14,* main topmast staysail; *15,* mizzen-topmast staysail; *16,* lower foretopsail; *17,* lower main topsail; *18,* lower mizzen-topsail; *19,* upper foretopsail; *20,* upper main topsail; *21,* upper mizzen-topsail; *22,* fore-topgallant sail; *23,* main topgallant sail; *24,* mizzen-topgallant sail; *25,* foreroyal; *26,* main royal; *27,* mizzen-royal; *28,* main skysail; *29,* main topgallant-staysail; *3o,* mizzen-topgallant-staysail; *31,* jib topsail; *32,* main topgallant sail; *33,* staysail; *34,* gaff-topsail.

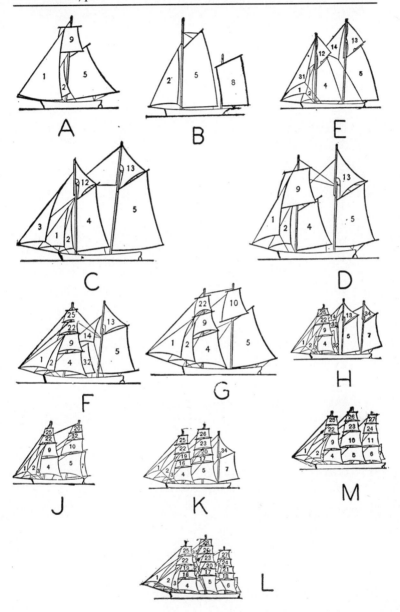

CLASS IX.   TRANSPORTATION

Section 56a.   Propellers and Paddle Wheels

A—Morgan's paddle wheel with feathering floats.

B—Feathering paddle-wheel buckets kept vertical by an eccentric ring.

C—Feathering paddle wheel.

D—Ericson's screw propeller; its rim is designed to counteract cavitation.

E—Griffith screw propeller with adjustable blade pitch.

F—Screw propeller with four blades.

G—Hodgson's screw propeller; its blandes curve backward.

H—Outward-thrust propeller wheel; its blades pitch forward.

J—Vergne's screw propeller; the projecting ribs serve to neutralize the centrifugal action of the water.

K—Reversing screw propeller.

L—Section view of the Voith-Schneider propeller; the constant speed makes possible a three-phase alternating current electrical drive with simple switch gear; the speed of the vessel is controlled by the variable pitch of the blades; steering is accomplished without a rudder.

M—Voith-Schneider propeller with vertical motor, may be used with both direct and alternating current.

N—Screw-propeller details.

O—Stern tube with lignum vitae bearings.

P—Modern ring-oiled propeller line shaft steady bearing.

Q—Grease-lubricated propeller line shaft steady bearing.

R—General arrangement of propeller stern tube and horseshoe-type thrust block.

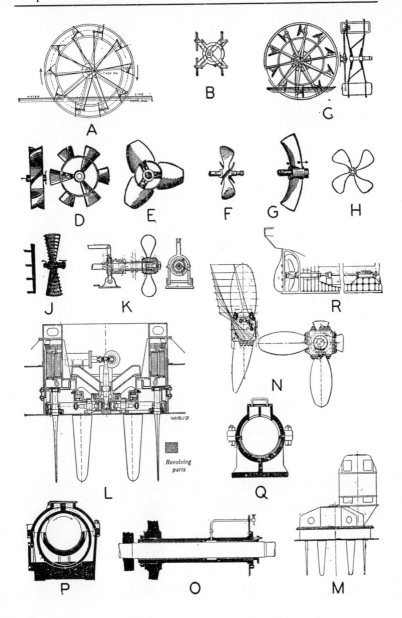

CLASS X.   INDUSTRIAL PROCESSES

Section 57a.   Weighing

A—Spring balance.

B—Circular spring-balance mechanism.

C—Small hand balance.

D—Original form of the Roberval balance; $h_1 = h_2$ irrespective of the horizontal positions of equal weights.

E—Roberval balance; diagram of forces acting on the leg.

F—Beranger balance.

G—Steelyard weighing machine.

H—Self-indicating weighing machine; in its simpler form, the scale is not linear, i.e., equal load increments are not indicated by equal divisions.

J—Self-indicating weighing machine with equal scale divisions for equal load increments; the tension rod is in the form of a flexible metal strip which operates on the periphery of a cam.

K—Laboratory balance (Welch Mfg. Co.).

L—Trip scale for rough weighing.

M—Dial scale for rough, quick weighing.

N—Prescription weights; made according to the specifications of the United States Bureau of Standards and the New York Board of Pharmacy; a set consists of 4, 2, 1 and $\frac{1}{2}$ drams; 2 and $\frac{1}{2}$ scruples; 10, 5, 2, 1, $\frac{5}{10}$, $\frac{2}{10}$ and $\frac{1}{10}$ grains.

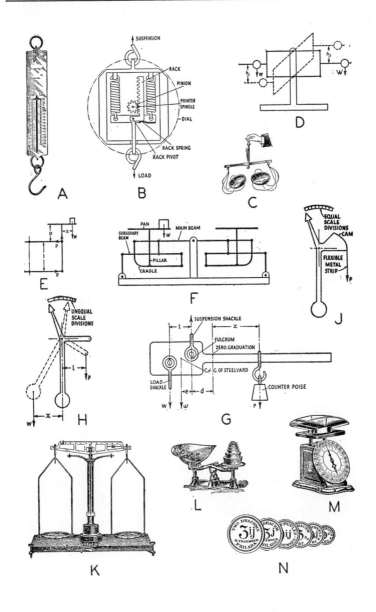

CLASS X.   INDUSTRIAL PROCESSES

Section 57b.   Measuring

A—Double-slide measure.

B—Automatic tipping scale; when it is full to weight, it falls and tips by striking a fixed stop; the scale then returns to its original position and is refilled.

C—Automatic measuring device; the material fills one compartment until it overbalances; then it falls and is emptied; the material then fills the other compartment; and the alternating motion continues while the feeding of material proceeds.

D—Measuring tap for liquids.

E—Automatic feeder.

F—Weighometer for automatically weighing any material passing along a horizontal belt, pan, or conveyor.

G—Internal micrometer and extension rod; measurements can be estimated to 0.0001 inch, a vernier scale being engraved on the barrel.

H—Solex air-operated gage; most suitable for the measurement of a large number of identical articles; it can be used as a thickness gage, a plug gage, or a ring gage; a special feature apart from its accuracy is the absence of wear on the gage unit; a common plug gage becomes useless after a certain number of passes, while a Solex plug gage is considerably smaller than the part to be measured, never comes in contact with part to be measured, and thus it is not subject to wear.

J—Outside caliper.

K—Inside caliper.

L—Vernier caliper.

M—Limit gage; go and not go snap gage.

N—Plug gage; one end is "go," the other end "no go."

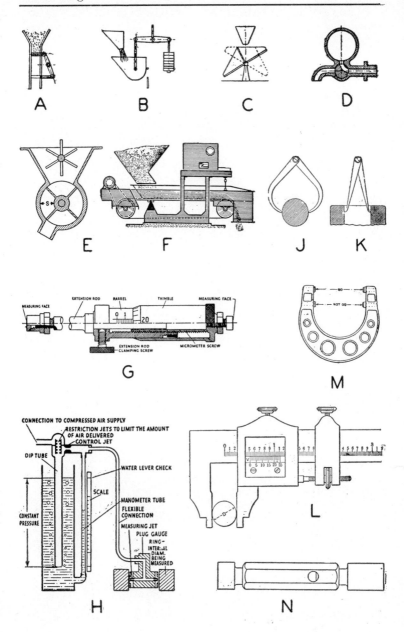

CLASS X.   INDUSTRIAL PROCESSES

Section 57c.   Pressure Measurements

A—U-tube or manometer for determining low gas pressures.

C—Bourdon tube-gage mechanism assembly.

D—Bourdon tube..

E—Diaphragm gage; the sensitive element is a corrugated disc of thin metal which, being held at the periphery, is deflected by a difference between the pressures on the two sides.

F—Reciprocating-engine or compressor indicator.

G—Metal bellows.

H—Aneroid barometer; vacuum chamber unit, showing diaphragms built up to form the unit.

J—Precision aneroid-barometer movement.

K—Sensitive altimeter; it is used to measure the height of an aircraft above ground level and also by surveyors.

L—Inclined draft gage.

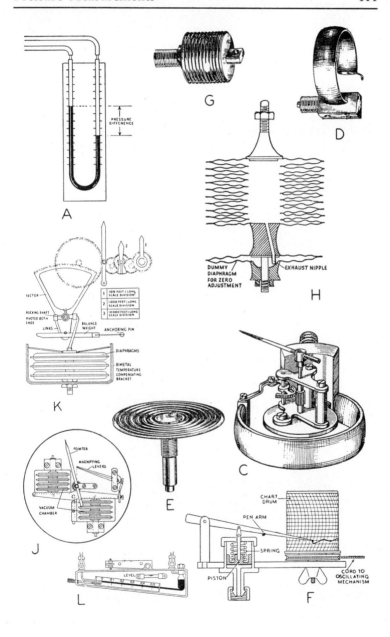

A

G

D

H

DUMMY
DIAPHRAGM
FOR ZERO
ADJUSTMENT

EXHAUST NIPPLE

SECTOR

1 | 100 FEET / LONG SCALE DIVISION
2 | 1.000 FEET / LONG SCALE DIVISION
3 | 10.000 FEET / LONG SCALE DIVISION

ROCKING SHAFT PIVOTED BOTH ENDS

LINKS

BALANCE WEIGHT

ANCHORING PIN

DIAPHRAGMS

BIMETAL TEMPERATURE COMPENSATING BRACKET

K

C

E

PRESSURE DIFFERENCE

POINTER

MAGNIFYING LEVERS

VACUUM CHAMBER

J

CHART DRUM

PEN ARM

SPRING

PISTON

CORD TO OSCILLATING MECHANISM

F

LEVEL

L

CLASS X.   INDUSTRIAL PROCESSES

Section 57d.   Speed and Heat Measurements

A—Principles of air-speed indicators.

B—Principles of the centrifugal speedometer.

C—Indicator to register the flow of water by its speed and pressure against a ball which actuates a pencil, moving against a paper cylinder which is kept slowly revolving by a clockwork.

D—Pitot tube; it measures velocity pressure and static pressure; the difference between the total pressure and static pressures gives the velocity pressure (VP).

E—Platinum resistance thermometer.

F—Wiring of a platinum resistance-thermometer bridge.

G—Beckman thermometer; its sensitivity is about 0.001°C; its range is extremely limited, about 1°C, but the position of this 1°C range can be varied at will over a wide range.

H—Wiring of a thermocouple; A and B are two dissimilar metals which are joined at their ends to form a circuit and if the junctions are kept at different temperatures, an electric current will flow in the circuit which can be measured by a galvanometer; one junction is placed in melting ice and the other in the temperature to be measured; up to 300°C, a copper-constantan couple is fairly sensitive; up to 1500°C, platinum-platinum rhodium is a satisfactory couple.

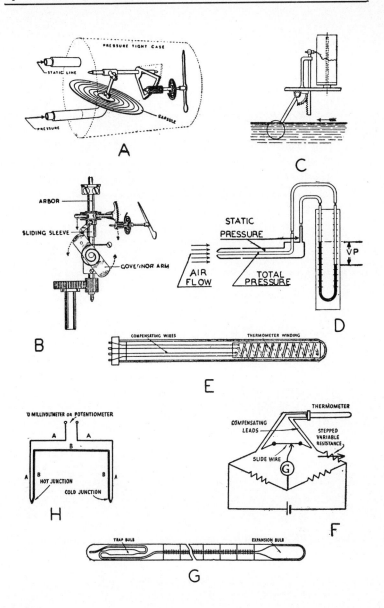

CLASS X. INDUSTRIAL PROCESSES

Section 58a. Crushing and Grinding

A—Conical mill and loop classifier; the material is fed into the mill by a constant-weight feeder; the ground material is removed from the mill as soon as ready; *3* separations from 20 mesh to as fine as 99.9%-325 mesh; *4* oversize particles are returned to the feed end of the mill; *5* the product is discharged at any elevation; *6* the fan returns clean air to the mill; *7* small amounts of air and vapor are vented; *8* "electric car" sound-control unit maintains maximum capacity (Hardinge).

B—Stone breaker with toggle motion.

C—Horizontal cone-plate mill.

D—Vertical cone mill.

E—Revolving stamp and pan mill for ores.

F—Revolving pan and ball mill.

G—Vertical cone grinding and crushing mill.

H—Crushing rollers with spring bearings.

J, K—Eccentric-disc grinding mill.

L—Planishing discs for rounding iron bars.

M—Ball mill for grinding various materials.

N—Cylindrical ball mill.

O—Huntington's stamp mill.

P—Grinding-face tool.

Q—Concrete mixer.

R—Device for axial rolling.

S—Lens grinder.

T—Cone-disc mill.

U—Cradle and roller mill.

V—Oscillating mill.

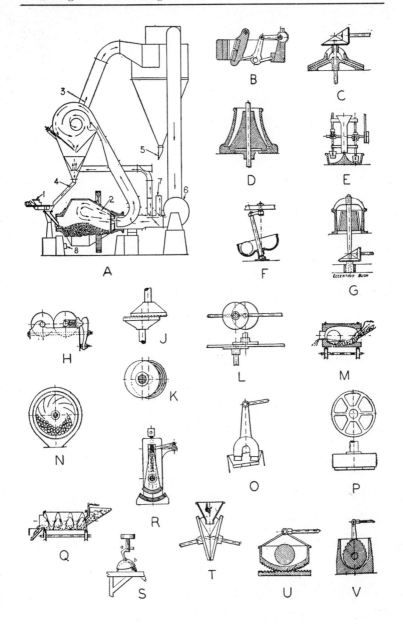

CLASS X.   INDUSTRIAL PROCESSES

Section 58b.   Crushing and Grinding

A—Stone breaker with chilled-iron jaw faces and toggle motion.

B—Horizontal centrifugal roller mill.

C—Cone-roller mill.

D—Stamp mill.

E—Cone-roller mill.

F—Lucop's patent centrifugal pulverizer.

G—Carr's patent disintegrator; each ring of bars is driven at high speed in opposite directions.

H—Double-edge runners; in some types the rollers revolve, in others, the pan revolves.

J—Enclosed cone-roller mill.

K—Conical-edge runner and pan.

L—Toothed-sector mill.

M—Rattle barrel for cleaning and polishing.

N—Ordinary flour mill; the grain is fed in the center, passes between the stones, and falls out into the outer casing.

O—Ball and pan mill for crushing ores.

P—Inclined ball and pan mill.

Q—Drum and roller revolving mill.

R—Single-roll coal crusher for chain-grate stokers.

S—Two-roll coal crusher.

T—Four-roll coal crusher (Bartlett-Snow).

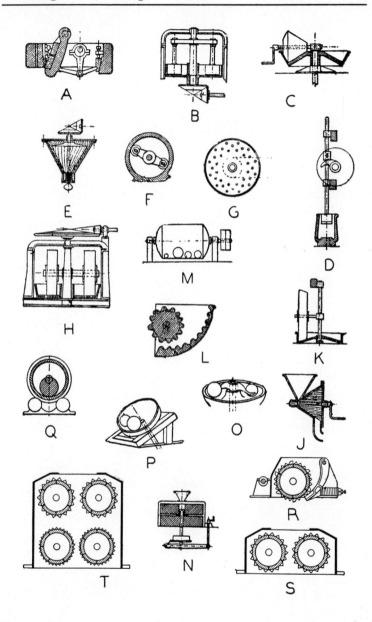

CLASS X.   INDUSTRIAL PROCESSES

Section 58c.   Sifting, Screening, Straining

A—Square-mesh wire gauze.

B—Perforated plate with round holes.

C—Parallel bars or wires.

D—Triangular mesh.

E—Hexagonal mesh.

F—Slit-end square-hole perforated plate.

G—Sloping screen.

H—Cylindrical- or slope-reel screen.

J—Cylindrical graduated screen or sizer.

K—Rotary screen, with rolling bevel-gear motion.

L—Shaking or jigging screen; it is sometimes supplied with a blast or aspirator to carry off the lighter particles.

M—Rotary horizontal screen.

N—Eccentric or angular barrel screen or mixer.

O—Air-blast sizing or graduating apparatus.

P—Edison's magnetic sizing apparatus for iron or steel particles.

Q—Graduating or sizing screens; fixed, or jigged like L.

R—Straight through pressure and suction strainer; the cover is removable for lifting out the perforated basket.

S—Cotton gin; *D* nest of saws; *E* saw grate between each saw to hold back the seed; *A* feeder trough and hopper; *J* cylinder brush stripping the cotton fiber from the saw; *F* adjusting lever; *K* sliding mote board.

T—Cyclone collector and separator; wood shavings or other light materials enter the inlet to the fan under suction and discharge into the cyclone at *1;* the material rotates around the periphery of the cylinder, drops into the cone, and is discharged at *3;* air is vented at *2.*

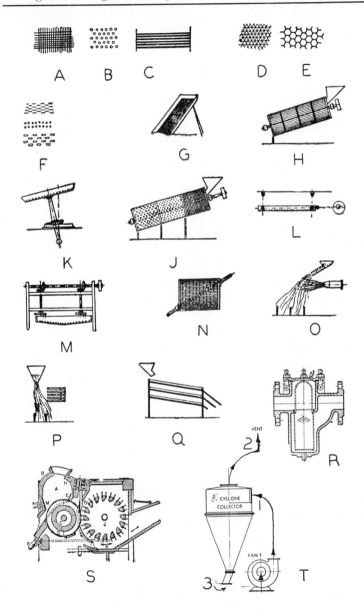

CLASS X.   INDUSTRIAL PROCESSES

Section 58d.   Chopping, Slicing and Mincing Apparatus

A—Wire-bending and cutting machine.

B—Disc cutter with radial knives and slots; it is used for vegetable fibers, roots, stems, etc.

C—Revolving cutter rollers.

D—Disc cutter with small knives wedged in separate holes through which cuttings escape in shreds.

E—Single-roller revolving cutter.

F—Spiral tapered revolving cutter; common mincing machine.

G—Revolving spiral cutter; lawn mower.

H—Two or more rectangular cutters, with vertical reciprocating motion in a revolving pan for mincing.

J—Apple slicer and corer.

K—Machine for slicing roots, etc.

L—Mill for chopping or grinding; it contains two rollers driven at different peripheral speeds.

M—Hand mincing compound knife.

N—Bark or cob mill.

O—Disc shears.

P—Hollow chuck with radial knives for rounding off wooden rods.

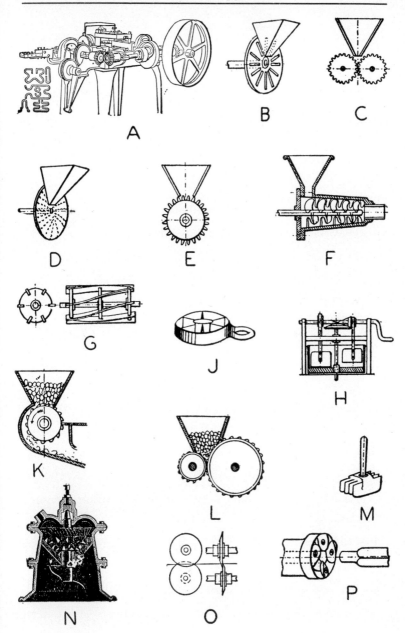

A

B

C

D

E

F

G

J

H

K

L

M

N

O

P

CLASS X. INDUSTRIAL PROCESSES

Section 58e. Mixers

A—Candy mixer operated by a crank.

B—Kneading mill with spiral vanes.

C—Pan mixer.

D—Mixing machine.

E—Dough mixer, or kneading machine.

F—Diagonal mixing pan used in confectionery.

G—Diagonal mixing barrel with fixed and revolving vanes

H—Mixer with two pairs of arms running in opposite di rections; the centers of the arms are above one another so that the arms pass each other in revolving.

J—Egg beater.

K—Horizontal table mixing machine; the material works its way from the center to the edge by centrifugal force.

L—Mixer with two pairs of arms running in opposite direc- tions.

M—Conical mixing barrel.

N—Pug mill with spiral paddles.

O—Pug mill with radial paddles revolving inside a conical case.

P—Concrete mixer.

Q—Circulating and screw-impeller mixer.

R—Vertical brine agitator in an ice-making tank.

S—Flash mixer for rapid mixing of chemicals (Link-Belt).

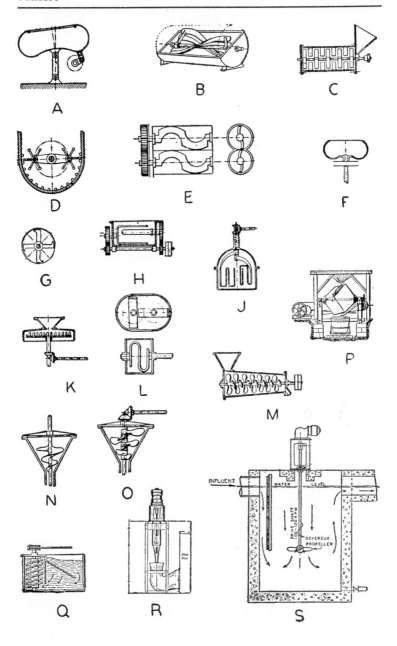

A

B

C

D

E

F

G

H

J

K

L

M

N

O

P

Q

R

S

CLASS X.   INDUSTRIAL PROCESSES

Section 58f.   Washing and Winding Apparatus

A—Revolving cylindrical screen washer.

B—Coal or ore washer.

C—Archimedean circulator for a washing trough.

D—Tub and paddle washer.

E—Cylindrical perforated drum with internal fixed spiral flange which causes the material to travel at a fixed speed; the cylinder may be revolved in a water trough as A, or water may be fed in with the material when the casing is unperforated.

F—Washing device for fabrics with corrugated roller.

G—Continuous circulator for boiling tubs; the hot water circulates similarly to the operation of a coffee percolator.

H—Rotary clothes washer; it consists of an internal perforated drum turning in alternating directions inside a vessel containing soap and water.

J—Water trough and dipping mechanism for washing paper fabrics.

K—Drum or barrel for winding wire, rope, etc.

L—Winding barrel for winches, cranes, etc.

M—Fusee barrel; it is used to give variable speed.

N—Grooved barrel for winding chain.

O—Hexagon-frame winder for chain.

P—Spool.

Q—Cord winder.

R, S—Bobbins.

T—Bobbin winder for cotton.

U—Drum for flat rope.

V—Fusee for round rope.

W—Thread feeding by an oscillating arm.

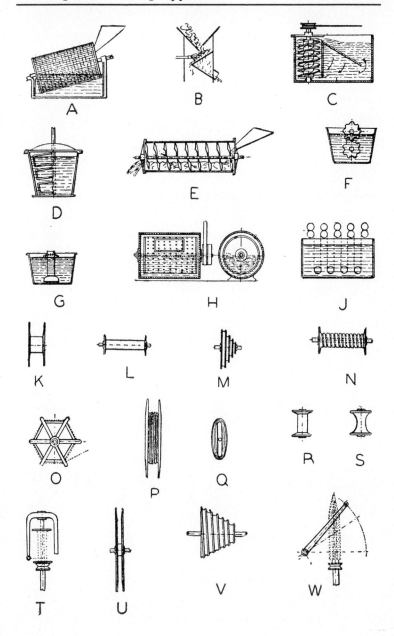

A

B

C

D

E

F

G

H

J

K

L

M

N

O

P

Q

R

S

T

U

V

W

CLASS X.   INDUSTRIAL PROCESSES

Section 59a.   Smithing and Forging

A—Cast-iron blacksmiths' hearth.

B—Brick-built blacksmiths' hearth.

C—Blast fan.

D—English anvil.

E—French anvil.

F—Blacksmiths' tools.

G—Swage block.

H—Dolly.

J—Arrangement of an industrial blacksmith shop.

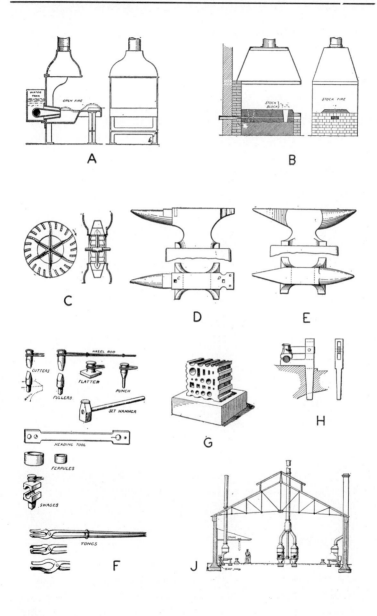

CLASS X.   INDUSTRIAL PROCESSES

Section 60a.   Presses

A—Rack and screw press.

B—Power press or stamp, with double crank movement.

C—Dick's antifriction press, with rolling contacts through-out.

D—Wedge press.

E—Screw fly press.

F—Revolving toggle press.

G—Sector and link press for increasing the pressure.

H—The "Boomer" double-toggle press with the pressure increasing as the follower descends.

J—Screw and toggle press.

K—Double-ram hydraulic press.

L—Lever press for hay and straw; the rack and pawl at each side are operated by two hand levers.

M—Continuous press for coal dust, etc.; the ram has a reciprocating motion; the material is forced into a tapered chamber.

N—Double-screw rubber vulcanizing press.

O—Cotton bale press.

P—Pascal's hydraulic press.

Q—Hydraulic press; with dies for lead pipe making; a similar press is used for making clay drain and flue pipes, the material being forced out of an annular orifice.

R—Ster-hydraulic press; a strand or rope is wound on a barrel inside the cylinder thus displacing water and raising the ram.

S—Press dies with a sliding plate for discharging.

T—Combined screw and hydraulic press; the screw is worked down by hand until the pressure becomes too high, then the pressing is finished by the hydraulic ram.

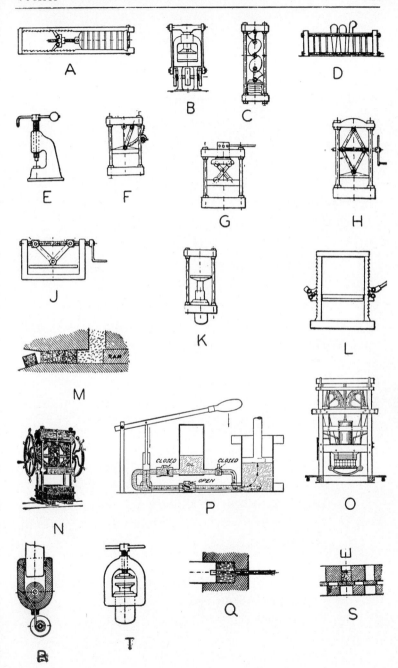

CLASS X.   INDUSTRIAL PROCESSES

Section 60b.   Drilling and Boring

A—V-drill for metal work.

B—Flat-point or bottoming drill.

C, D—Countersinking drills for metals.

E—Center bit for wood.

F—Twist bit for wood; it clears its own borings.

G—Earth borer or mooring screw.

H, J, K—Rock drills or jumper.

L—Twist drill for metals.

M, N—Countersinking drills for wood or metal.

O—Diamond rock drill; it bores an annular hole, the core of which cracks off at intervals and passes up the tube.

P to Z—Well-boring tools for different kinds of strata; tools for raising broken rods, etc.

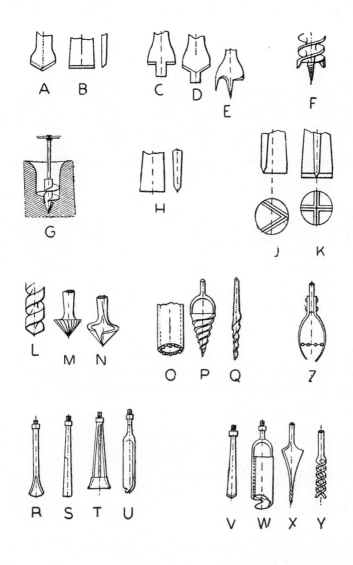

CLASS X.   INDUSTRIAL PROCESSES

Section 60c.   Cutting Tools

A—Pipe cutter with V-edged cutting roller.

B—Slitting discs for cutting sheets into strips.

C—Revolving cutter head, for tenoning, molding, etc.

D—Reaping machine cutters with a series of scissor-shaped knives, one set fixed and the other reciprocating.

E—Wire-cutter discs, one fixed, the other attached to a hand lever; they have corresponding holes of various sizes in both discs.

F—Guillotine sheers.

G—Tubular machine cutter for woodworking; it is easily sharpened and can be revolved to present fresh cutting edges to the work.

H—Scroll saw, fret saw, or jigger.

J—Three-cutter tube shears with worm-gear motion.

K—Pin borer for cutting out circular blanks with a central hole, such as washers, etc.

L—Inserted circular saw teeth easily sharpened or replaced.

M—Expansive facing, or boring pin bit.

N—Revolving cutter, with adjustable inserted circular cutter.

O—Wobbling circular saw for cutting dovetail grooves.

P—Chain cutter.

Q—Turning knife tool for metal cutting.

R—Turning front tool for metal cutting.

S—Chasing tool for cutting V-threads.

T—Hollow-taper bung borer.

U—Screw tool for cutting square threads.

V—Cutting discs for sheet metal, paper, etc.

W—Compound lever shears.

X—Square-hole boring bit for wood; a square chisel containing a twist drill.

Y, Z—Cylinder and fluted drills for enlarging and finishing holes.

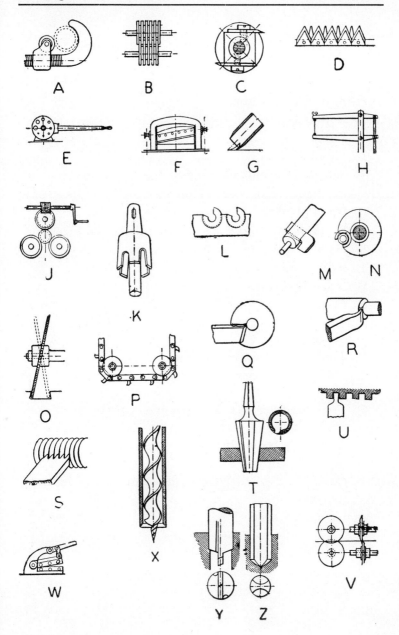

A

B

C

D

E

F

G

H

J

K

L

M

N

O

P

Q

R

S

T

U

W

X

Y

Z

V

CLASS X.   INDUSTRIAL PROCESSES

Section 60d.   Cutting Tools

A—Right-hand turning tool for metal.

B—Boring tool.

C—V-tool.

D—V-tool for inside threads.

E—Side tool for square shoulders.

F, G—Boring tools for square shoulders.

H—Hand planing tool for soft metals, e.g., lead.

J—Hand planing tool for wood with the grain.

K—Hand planing tool for end grain.

L—Paring gouge for wood.

M—Hollowing gouge for wood.

N—Cross grooving plane; it has two cutters, one to mark the cut on each side, the other to plane out the shaving.

O—Tool for cutting circular holes in a steel plate.

P—Hollow-cone paring tool for pointing pins or lead pencils.

Q—Tool head for drilling machine with three or four drill holders.

R, S—Adjustable boring bits.

T, U—Bottoming or rose drills.

V—Compound cylinder drill; fluted and provided with coolant channel.

W—Boring-bar head.

X—Tool holder for a lathe, shaper, or planer.

Y—Coal-cutting wheel.

Z—Tunnel-boring head for clay or soft soils.

AA—Coal-cutting borer.

BB—Broach.

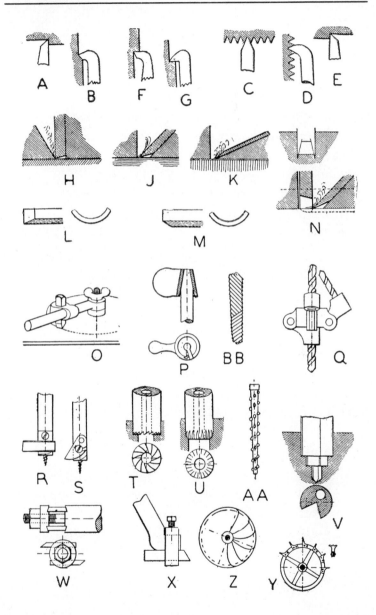

CLASS X.   INDUSTRIAL PROCESSES

Section 60e.   Cutting Tools

A, B—Boring bar.

   C—Plain milling cutter with spiral teeth.

   D—Plain milling cutter with straight teeth.

   E—Solid milling cutter with rake teeth for slotting, straddle milling, or general side or face milling.

   F—Spur-gear hob for cutting gear teeth.

   G—Inserted-blade milling cutter, using tool steel, stellite, or carbide-tipped inserts.

   H—Gear cutter with radial teeth.

   J—Multiple-thread miller for cutting threads.

   K—Slitting saw with rake teeth for milling deep slots.

   L—End mill with spiral teeth for die sinking, and for milling surfaces not conveniently reached by other cutters.

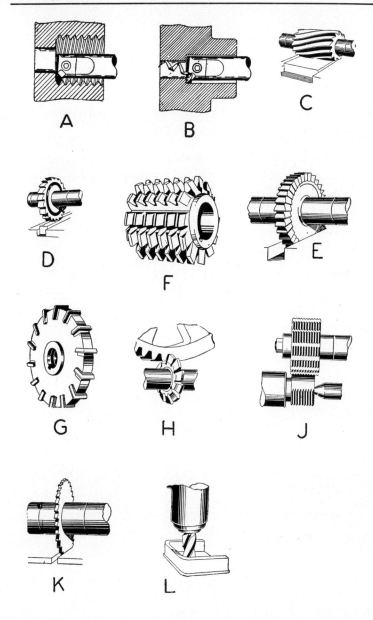

A

B

C

D

F

E

G

H

J

K

L

CLASS X.   INDUSTRIAL PROCESSES

Section 60f.   Cutting Tools

A—Hydraulic cold saw; it is driven by a separate motor and is fed into the metal by hydraulic pressure.

B—Hydraulic metal saw; hydraulic power clamps the work and also actuates the feed of the saw as it cuts its way through the work.

C—Oxy-acetylene torch for cutting steel.

D—Teeth of a cross-cut wood saw.

E—Teeth of a rip wood saw.

F—Teeth of rip saw.

G—Side view of the teeth of a rip saw.

H—Hack saw for metals.

J—Back saw.

K—Compass saw.

L—Flooring saw.

M—Pattern maker's saw.

N—Dovetail saw.

O—Keyhole saw.

P—Bench or joiner saw.

Q—Stanley block plane for planing across the grain; the angle is 20° for soft wood.

R—Quarter round molding plane.

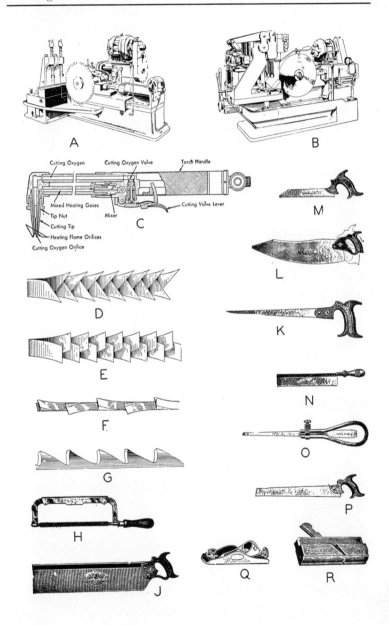

A

B

Cutting Oxygen   Cutting Oxygen Valve   Torch Handle

Mixed Heating Gases

Cutting Valve Lever

Tip Nut   Mixer

Cutting Tip   C

Heating Flame Orifices

Cutting Oxygen Orifice

M

L

D

K

E

N

F

O

G

P

H

J   Q   R

CLASS X.   INDUSTRIAL PROCESSES

Section 61a.   Agricultural Machinery

A—Spanish plow (1708).

B—Daniel Webster's plow (1836).

C—Chilled plow with reversible point; made of cast iron; designed by Oliver.

D—Sulky plow (1910).

E—Reversible or hillside plow, for use where the land is too sloping to throw the furrow-slice uphill.

F—Modern disc plow for horse or tractor power.

G—Disc engine gang.

H—Smoothing harrow with wooden frame.

J—Weeder for killing young weeds.

K—Curved knife-tooth riding harrow, clod crusher and leveller.

L—Spading-disc harrow.

M—Orchard-disc harrow; it has a wide frame for work under trees.

N—Smooth iron roller.

O—Corrugated roller.

P—Riding cultivator with eight shovels; it has a hammock seat and balance frame.

Q—Disc cultivator.

R—Original Campbell combined cultivator and grain drill

S—Potato planter in cross-section.

T—Potato digger; high-elevator type.

U—Potato digger; low-elevator type.

CLASS X.   INDUSTRIAL PROCESSES

Section 62a.   Sanitation; Water Closets and Septic Tanks

A–Combination of venting and sewage discharge permitted under certain conditions by the United States Government; the soil stack is at one side of the battery of fixtures and the vent on the other, while the horizontal line, to which the fixtures are connected, acts as a combination vent and soil or drain line; this installation has a weak point as the branches from the individual fixtures to the combination horizontal pipe must be over 2 feet long due to construction, and when they are not supplied with an individual vent, they are similar in location to a trap on the highest fixture in a building with a branch over 2 feet long without an individual vent which causes self-siphonage; this combination is not allowed by the New York City code nor is it by others.

B–Loop venting; a stoppage between the upstream fixtures and the ones below prevents air from entering the pipe, causing unbalanced pressure and siphonage.

C–Individual venting of each trap; the sure way to prevent the siphonage of fixtures.

D, E–Design for duplex bathroom group; note incorrect connection; care must be taken not to make the branch to the fixture over 24 inches or to have the end of the branch lower than the bottom of the trap, otherwise, the trap will siphon; the New York code requires two separate vents for the water closets.

F, G–Septic tank with siphon chamber; concrete construction; capacity for residences 60 gallons per person; depth of liquid about 5 feet.

H, J–Simple septic tank; concrete construction; for day schools; 15 gallons per person generally allowed.

K–Septic privy; "Kentucky" sanitary privy; suitable for ten to twelve persons.

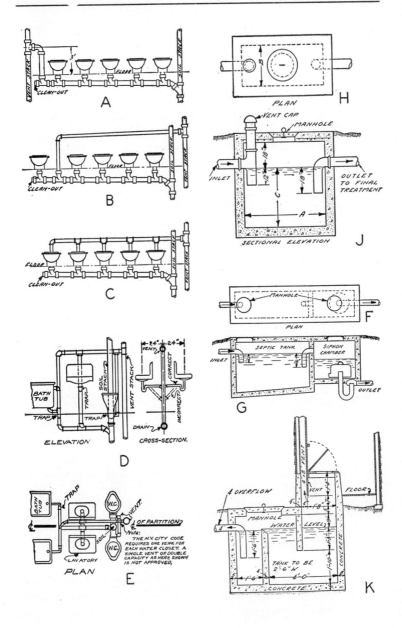

CLASS XI. ELECTRICAL APPLIANCES

Section 63a.   Static Electricity

A—Positive and negative electricity; when glass is rubbed with silk, it becomes positively charged; when sealing wax or gutta percha is rubbed with flannel, it becomes negatively charged; like charges repel, unlike attract each other.

B—Electrophorus; it consists of a hard-rubber plate *P*, which can be electrified by friction and a metal disc *C*, which has an insulated handle *H*, and which is called a carrier; the hard-rubber plate *P* generally rests on a metal plate *S* which is called the sole plate; rubbing the plate with fur charges it negatively.

C—Electroscope; used for detection of an electric charge; it consists of a metal rod supported by a stopper of sealing wax or sulfur and carrying two leaves of aluminum or gold foil suspended from its lower end, the leaves and rod being enclosed in a glass flask, if an electrified body is brought into contact with the metal ball on the end of the rod, the gold or aluminum leaves will diverge, since they become similarly charged and repel each other, the amount of repulsion being a measure of the strength of the charge.

D, E, F—Charging the electroscope; *D*, to charge it negatively by induction, bring a positively charged rod near the knob, the leaves diverge; *E*, holding the rod near the knob, touch the knob with the finger and the leaves collapse; *F*, withdraw first the finger, then the rod and the leaves diverge, being negatively charged.

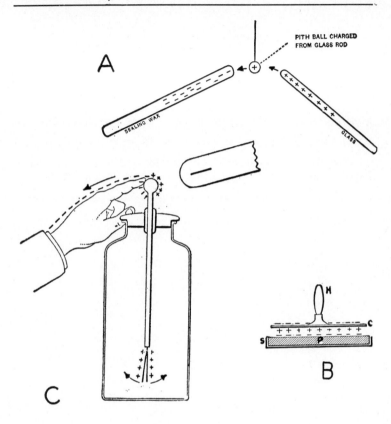

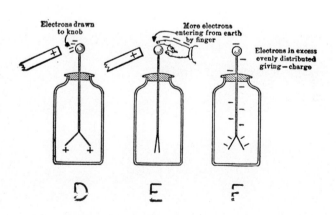

CLASS XI. ELECTRICAL APPLIANCES

Section 63b.   Static Electricity

A—Proof plane; used when testing a body for electrification, which is too heavy or inconvenient to carry to the electroscope; a copper penny is waxed to the end of a glass or rubber rod; by touching the penny alternately to the body and to the electroscope, it can be determined whether or not the body is charged.

B—Charging by induction; a body may acquire an electrical charge by contact with another charged body.

C—Electric field around a sphere.

D—Electric charge accumulates on the outside of a hollow conductor.

E—Electric charge accumulates near the pointed end of a body.

F—Electroscope measuring potential.

G—Plate condenser.

H—Condenser made of two parallel plates.

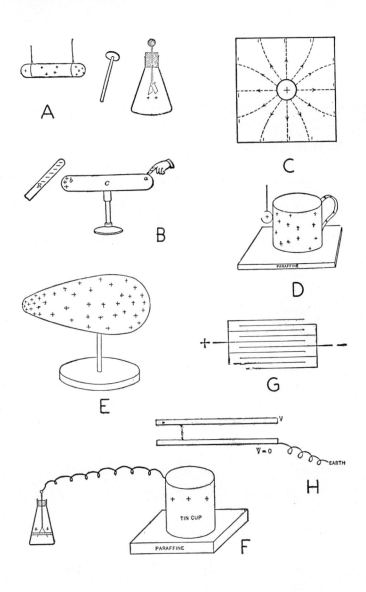

CLASS XI.   ELECTRICAL APPLIANCES

Section 63c.   Magnetism

A—Magnetic compasses.

B—Magnet attracting iron filings.

C, D—Arrangements of magnetic domains in an iron bar: C before magnetization; D after magnetization.

E—Magnetic induction with and without contact.

F—Magnetic field about a bar magnet.

G—Magnetic fields around two bar magnets showing attraction.

H—Magnetic fields around two bar magnets showing repulsion.

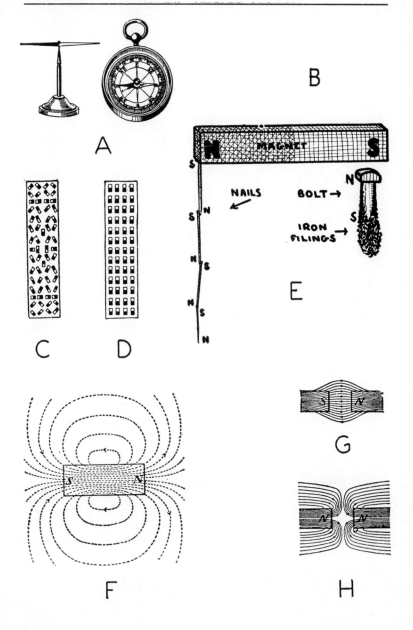

A

B

C

D

E

NAILS

BOLT →

IRON FILINGS →

F

G

H

CLASS XI. ELECTRICAL APPLIANCES

Section 63d. Electromagnets

A–Simple electromagnet.

B–Iron filings in a magnetic field about a conductor carrying a current.

C–Right-hand thumb rule; thumb shows direction of current; fingers show direction of magnetic lines of force.

D–Reversed right-hand thumb rule; fingers show direction of current in the turns of wire and the thumb indicates the north pole of the electromagnet.

E–Horseshoe electromagnet; coils of insulated wire around a soft iron core; an electric current flowing through the wire forms a magnet of greatly increased strength.

F–Electric bell; the armature is fastened to a spring which is so adjusted that its tension is sufficient to maintain the armature in contact with a screw when no current is flowing; when the button is pushed, the electromagnet is energized and attracts the armature, thus ringing the bell; as the armature moves toward the magnet, the circuit is broken; the spring pulls the armature back to make contact with the screw and the process is repeated.

G–Electric-bell circuit.

H–Telegraph key; it consists of a lever and two contact points; when the lever is pressed, the circuit is closed; when the lever is released, the circuit is broken.

J–Sounder; a device for making clicks; when the key is pressed, the circuit is completed and the electromagnet attracts the armature and *H* is struck, making a clicking sound; when the key is raised, the circuit is broken, the armature is released, and the spring pulls the lever *B* up, making another click.

K–Telegraph relay; it consists of an electromagnet which responds to weak electric impulses and passes messages on to a strong local current which operates a sounder.

L–Complete telegraph system.

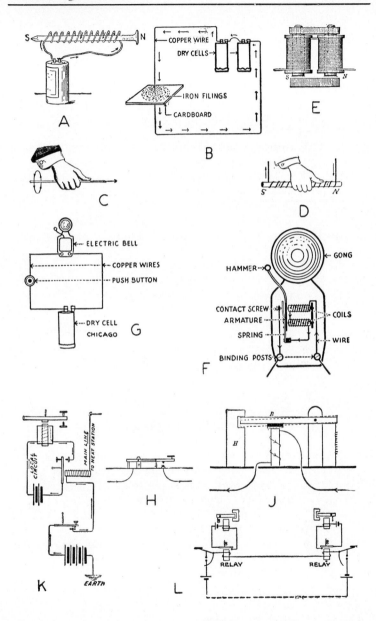

A

B

← COPPER WIRE
DRY CELLS →
IRON FILINGS
CARDBOARD

E

C

D

ELECTRIC BELL
COPPER WIRES
PUSH BUTTON
DRY CELL
CHICAGO

G

GONG
HAMMER
CONTACT SCREW
ARMATURE
SPRING
BINDING POSTS
COILS
WIRE

F

MAIN LINE
TO NEXT STATION
LOCAL CIRCUIT

K

EARTH

H

J

L

RELAY   RELAY

CLASS XI.   ELECTRICAL APPLIANCES

Section 63e.   Basic Types of Permanent Magnets*

A–Simple bar magnet.

B–U-shape magnet.

C, D–Other arrangements of two poles to produce similar results to B.

E–Soft-steel plates are north or south around the entire periphery; this arrangement may be used as a roll for handling sheet metal.

F–Simple air gap made to definite dimensions.

G, H, J–Simple air gap made to definite dimension with soft-steel pole pieces.

K, L, M–Soft-steel return circuit.

N, O, P–Magnet surrounding an air gap; used where the available space is limited.

Q–Two magnets placed together to form a double air gap, the direction of flux being opposite in each air gap.

R–Air gap at either end.

S–Annular air gap.

T, U, V–Other annular combinations.

W–Rotating magnet with two poles.

X, Y, Z–Built-up rotors.

AA–Four-pole rotating magnet; it is often desirable to use soft-steel poles on magnets of this type.

BB, CC, DD–Built-up four-pole rotors with soft-steel poles.

EE–Cylindrical magnet magnetized with poles about the periphery; this is possible only with modern permanent-magnet materials.

FF–A variation of EE.

GG–Poles on the inside of a ring.

HH, JJ, KK–Magnets which can be made with any desired number of poles.

LL to XX–Fundamental types of air gaps.

*Courtesy of Indiana Steel Products Co.

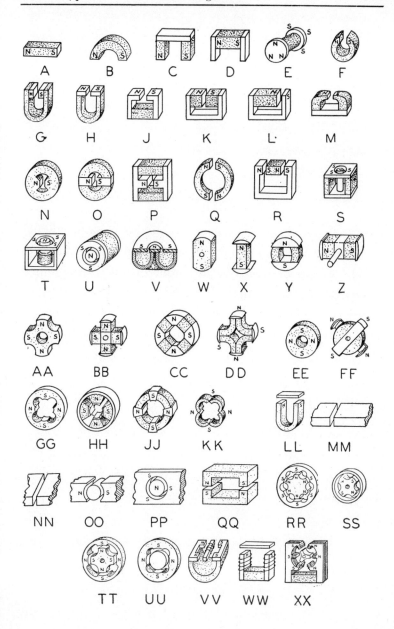

A　　B　　C　　D　　E　　F

G　　H　　J　　K　　L　　M

N　　O　　P　　Q　　R　　S

T　　U　　V　　W　　X　　Y　　Z

AA　　BB　　CC　　DD　　EE　　FF

GG　　HH　　JJ　　KK　　LL　　MM

NN　　OO　　PP　　QQ　　RR　　SS

TT　　UU　　VV　　WW　　XX

CLASS XI.   ELECTRICAL APPLIANCES

Section 63f.   Batteries

A—Various parts of a voltaic cell.

B—Simple voltaic cell.

C—Local action; commercial zinc contains metallic impurities and when placed in an acid, small electric currents are set up between two different metals, causing local action or corrosion.

D—Polarization; a defect caused by bubbles of hydrogen gas accumulating on the positive plate of a battery.

E—Gravity cell.

F—Weston standard cell; it produces a constant electromotive force and is used as a standard of measurement.

G—Leclanché dry cell.

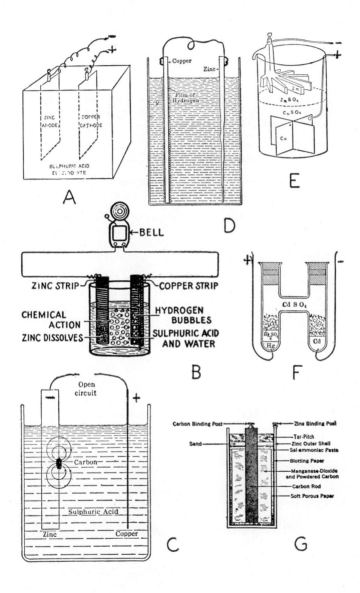

A

B

C

D

E

F

G

CLASS XI.   ELECTRICAL APPLIANCES

Section 63g.   Chemical Effects of Electricity

A—Electrolysis of water; when a direct current flows through a solution of sulfuric acid diluted with water, hydrogen accumulates at the cathode and oxygen at the anode.

B—Copper plating; the object to be plated is connected to the cathode and a plate of copper serves as the anode.

C—Charging a storage battery; the positive plates become coated with lead peroxide; the negative plate becomes spongy lead.

D—A storage battery discharging direct current; the positive plate releases oxygen, lead peroxide changing to lead sulfate; the negative plate becomes coated with lead sulfate; the sulfuric acid becomes more dilute.

E—Electrolytic copper refining; impure copper is used as positive electrodes and strips of pure copper as negative electrodes; when a direct current is passed through the cell, copper of the positive electrodes goes into solution and pure copper is deposited on the negative electrodes; impurities, including gold and silver, settle to the bottom.

F—Charging a battery with constant 110-volt direct current; several batteries may be connected in series; the lamp bank offers series resistance.

G—Silver plating; it is similar to copper plating; the solution consists of silver nitrate and potassium cyanide; the object to be plated and a strip of pure silver are dipped into the solution and connected to a current supply; the object to be plated is charged negatively and the silver sheet positively.

H—Downs cell for making sodium from sodium chloride; the positive electrode made of graphite projects through an iron box lined with fire brick; the negative electrode is a band of iron or copper which encircles the graphite separated by a wire gauze; sodium is produced from the fused salt (600°F) at the negative electrode and flows into the receiver *C;* chlorine, a valuable by-product, is collected at *B;* sodium chloride is added from time to time at *A* and is kept in a molten state by its resistance to the flow of the electric current.

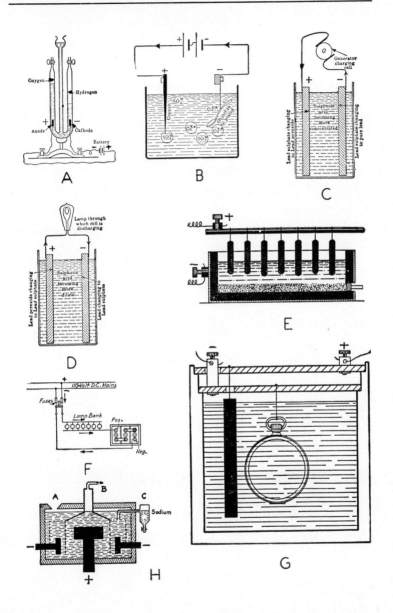

CLASS XI.   ELECTRICAL APPLIANCES

Section 63h.   Transformers; Induction Coils

A—Simple transformer; the primary coil P consists of fine wire and the secondary coil S consists of heavy wire to induce a large alternating current of low voltage in S when a small current of high voltage passes through P.

B—Transformers used to transmit electric power over long distances.

C—Induction coil.

D—Detailed section of an induction coil.

E—V-connections for transforming moderate amounts of three-phase power.

F—T-connection for transforming moderate amounts of three-phase power.

G—T-connection for transforming from three-phase to two-phase power.

H—T-connection for transforming from three-phase to four phase power.

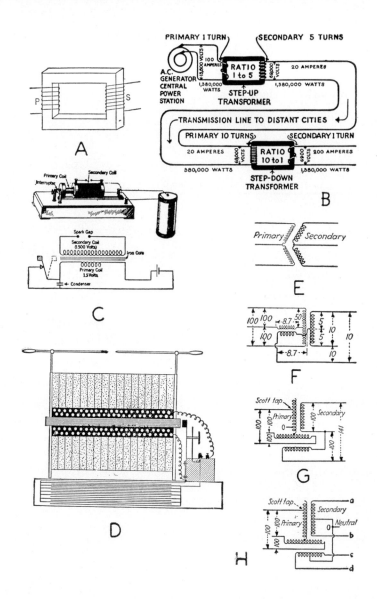

A

B

PRIMARY I TURN | SECONDARY 5 TURNS

A.C. GENERATOR CENTRAL POWER STATION | 100 AMPERES | 13,800 VOLTS | RATIO I to 5 | 69,000 VOLTS | 20 AMPERES

1,380,000 WATTS | STEP-UP TRANSFORMER | 1,380,000 WATTS

TRANSMISSION LINE TO DISTANT CITIES

PRIMARY IO TURNS | SECONDARY I TURN

20 AMPERES | 69,000 VOLTS | RATIO 10 to 1 | 6900 VOLTS | 200 AMPERES

380,000 WATTS | STEP-DOWN TRANSFORMER | 1,380,000 WATTS

Primary Coil | Secondary Coil | Interrupter

Spark Gap | Secondary Coil (1,500 Volts) | Iron Core | Primary Coil 1.5 Volts | Condenser

C

D

Primary | Secondary

E

100 100 | 8.7 | 50 | 5 5 10 | 100 | 10 | 8.7

F

Scott tap | Secondary | 100 100 | Primary | 0 | 100 141 | 100

G

Scott tap | Secondary | a | 100 | Primary | 0 Neutral | 100 | b | c | d

H

CLASS XI.   ELECTRICAL APPLIANCES

Section 63j.   Telephones

A—Simple, one-way telephone system; essential parts are a transmitter, a receiver, an induction coil, and an electric battery.

B—Transmitter; sound waves striking the flexible diaphragm push it in, which increases the pressure on the carbon granules thus reducing the electrical resistance; a refraction of the sound waves releases the diaphragm, decreases the pressure on the carbon, and increases the electrical resistance; the varying resistance causes continual fluctuations of current, which, acting on an induction coil, causes the diaphragm of receiver to vibrate in accordance with the vibrations of the transmitter.

C—Circuit at the receiving end.

D—Telephone receiver.

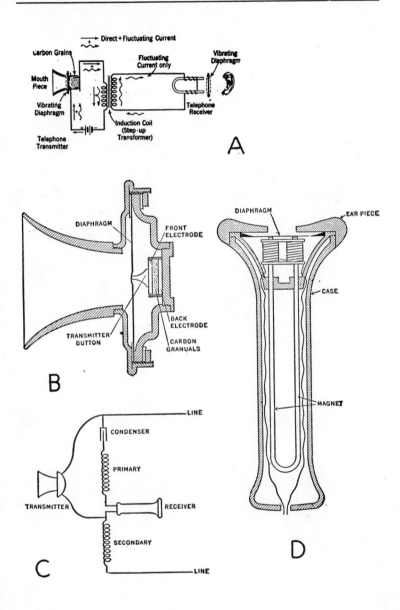

CLASS XI.   ELECTRICAL APPLIANCES

Section 63k.   Electrical Measuring Instruments

A—Voltmeter; used for measuring the difference of potential or electromotive force between two points in an electric circuit; r is a high-resistance coil.

B—D'Arsonval galvanometer; used to detect the presence of an electric current, its direction and strength.

C—Ammeter; used for measuring the strength of a current in amperes; t is a shunt.

D—Kelvin multicellular electrostatic voltmeter.

E—Leeds and Northrup reflecting galvanometer; its sensitivity permits detection of a current as weak as 0.0001 microamperes.

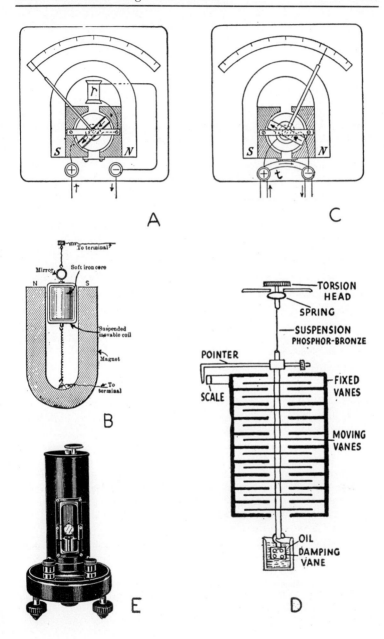

A

C

B

D

E

TORSION
HEAD
SPRING

SUSPENSION
PHOSPHOR-BRONZE

POINTER

SCALE

FIXED
VANES

MOVING
VANES

OIL
DAMPING
VANE

To terminal

Mirror   Soft iron core

N   S

Suspended
movable coil

Magnet

To
terminal

CLASS XI.   ELECTRICAL APPLIANCES

Section 631.   Electrical Measuring Instruments

A—Small voltaic cell and coil floating in a dish of water; when the north pole of a bar magnet is brought near the coil, the circuit will always rotate so as to present that face in which the current is flowing clockwise; i.e., the south face of the coil is attracted by the north pole of the magnet; also the coil will move so as to include more lines of force; this illustrates Maxwell's rule.

B—Tangent galvanometer (W. M. Welch Mfg. Co.).

C, D—Moving indicators; C shows gravity control; D illustrates spring control; the moving systems on the indicators are suspended or pivoted in jewelled bearings which can be twisted through an arc of 90 to 270°; the magnetic, thermal, or electrostatic effect is made to exert a torque; the angle of deflection is indicated by a pointer attached to the moving system or by a light spot reflected from a mirror, and is a measure of the current or voltage.

E—Piston damper; prevents undue oscillations of an indicator.

F—Vane damper.

G—Moving-coil permanent-magnet system.

H—Attraction-type moving iron instrument.

J—Repulsion-type moving iron instrument.

K—Galvanascope; the current-carrying part of the wire is movable and the magnet is fixed.

L—Therma-galvanometer; adapted to measure a small alternating current which changes direction so rapidly that it does not affect the coil of a galvanometer; the amount of heat depends on the square of the current and the heating effect is independent of the direction of the current; the heating effect offers a method of measuring alternating currents; antimony and bismuth form the thermocouple.

M—Thermocouple instrument; when the junction of two dissimilar metals is subjected to different temperatures, an electrical potential is set up between them which is proportional to the temperature difference; if a moving-coil ammeter is placed in the circuit, it will measure the current flowing when the temperature of one junction is raised by heating.

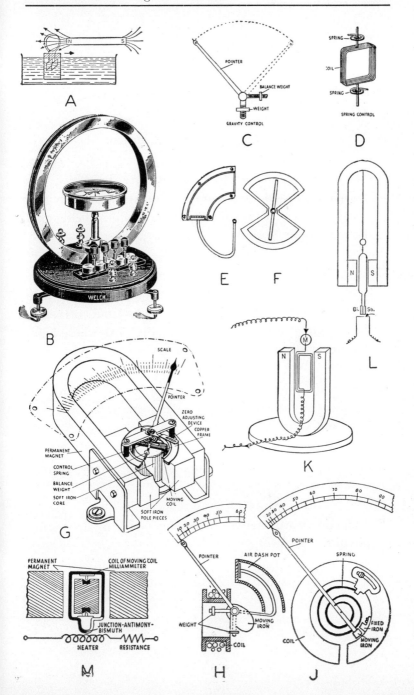

CLASS XI. ELECTRICAL APPLIANCES

Section 63m. Electrical Measuring Instruments

A—Weston direct-current movable-coil system.

B—Standard dynamometer-type wattmeter in which the permanent magnet with moving coil is replaced by two fixed coils which provide an almost uniform field; used for measuring power.

C—Hot-wire instrument for measuring alternating currents exclusively; it depends on the expansion of a wire which is heated by the current or part of the current to be measured; used principally in radio work where high-frequency currents are to be measured.

D—Induction instrument; used in alternating circuits, on frequencies for which they are designed, to measure current and voltage.

E—Shaded-pole induction instrument for measuring current and voltage.

F—Induction wattmeter for measuring power in alternating circuits.

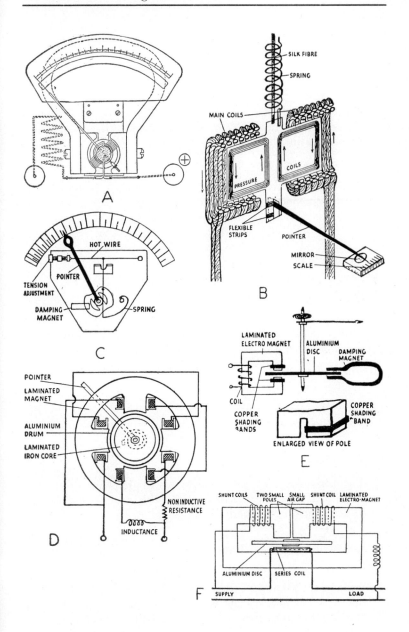

A

SILK FIBRE

SPRING

MAIN COILS

PRESSURE COILS

FLEXIBLE STRIPS

POINTER

MIRROR

SCALE

B

HOT WIRE

POINTER

TENSION ADJUSTMENT

DAMPING MAGNET

SPRING

C

LAMINATED ELECTRO MAGNET

ALUMINIUM DISC

DAMPING MAGNET

COIL

COPPER SHADING BANDS

COPPER SHADING BAND

ENLARGED VIEW OF POLE

E

POINTER

LAMINATED MAGNET

ALUMINIUM DRUM

LAMINATED IRON CORE

NON INDUCTIVE RESISTANCE

INDUCTANCE

D

SHUNT COILS   TWO SMALL POLES   SMALL AIR GAP   SHUNT COIL   LAMINATED ELECTRO-MAGNET

ALUMINIUM DISC   SERIES COIL

F

SUPPLY   LOAD

CLASS XI.   ELECTRICAL APPLIANCES

Section 63n.   Electrical Generators

A—Simple coil of wire carrying an electric current which produces a magnetic field; the field is stronger if a steel bar is placed within the coil to form a core.

B—Horseshoe core.

C—Current generation in an electric conductor by moving it through a magnetic field; the direction of the voltage depends on the direction of the field and on the direction of the motion of the conductor; the strength of the current depends on the speed of the conductor and the strength of the magnetic field.

D—Simple electric generator with single-phase, revolving armature.

E—Loop of wire rotating in a magnetic field; the generated voltage depends on the speed of cutting across the field, on the field strength, and on the length of the wire.

F—Single-phase voltage curve, representing variations in voltage during two revolutions in a magnetic field.

G—Two simple generators rotating on the same shaft; the single-loop armatures are always at right angles to each other and in fields of the same strength.

H—Two-phase voltage curve.

J—Two-phase alternator formed by two single-phase alternators combined into one machine with both armature windings rotating in one magnetic field.

K—Three separate loops of wire revolving in the same magnetic field with each loop brought out to a separate pair of slip rings and a separate external circuit.

L—Voltage variation of a three-phase alternator; at every point, the value of one voltage is equal and opposed to the sum of the other two, values below zero being negative.

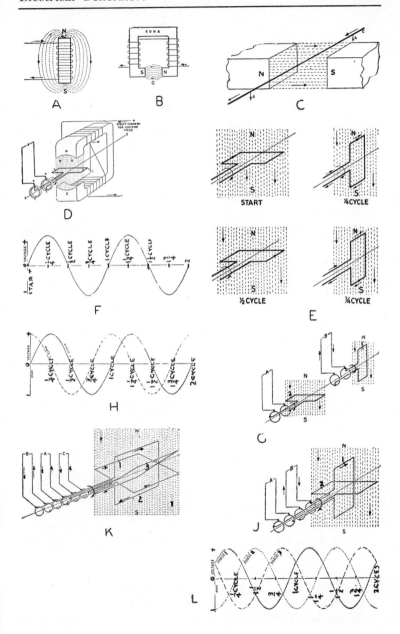

CLASS XI. ELECTRICAL APPLIANCES

Section 63o. Electrical Generators

A—Three loops connected together within the alternator; only three slip rings are required, because each ring can serve two outside circuits.

B—Transformer used for changing the voltage of alternating currents; the windings are insulated from each other.

C—Autotransformer having only one winding which is divided into two parts; it changes the voltage, but does not insulate the two circuits.

D—Relation between alternating and direct currents; a direct current has a steady value; an alternating current continually changes; the two currents shown have the same effective value.

E—Simple direct-current generator with two poles separately excited; a commutator is used to keep the current in the inside circuit flowing continually in the same direction.

F—Simple single-phase alternator with a revolving armature and four poles; the number of poles does not determine the number of phases.

G—Single-phase alternator with six poles in a revolving field.

H—Current, voltage, and power in a circuit operating at unity power factor; the current rises and falls with the voltage, and the power is always positive; at every point the product of current and voltage gives the power.

J—Current, voltage, and power in a circuit containing inductance; the current always lags behind the voltage so that the product of current and voltage is not always positive; the power is always zero when the current or voltage is zero.

K—Slip rings.

L—Split-ring commutator.

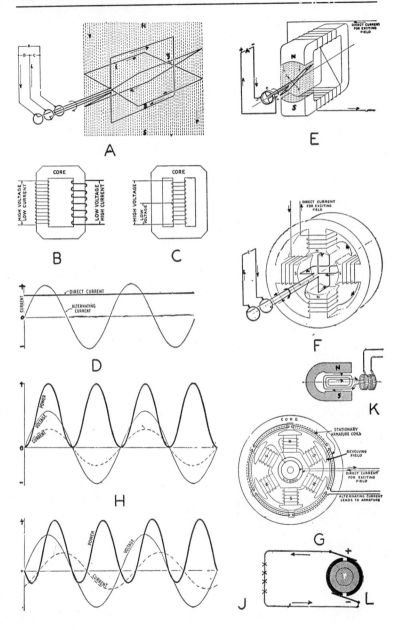

CLASS XI.   ELECTRICAL APPLIANCES

Section 63p.   Alternators

A—Polyphase alternator which may be phased out by means of lamps; if the phase rotation is correct on the incoming alternator, the lamps will be all dark or they will all light up simultaneously.

B—Checking of phase rotation by means of a small polyphase induction motor connected alternately to the bus and the incoming machine; if the direction of rotation of the motor is the same for both the incoming alternator and the bus, the phase rotation is the same; after the phase rotation has been once established and made correct, it is not necessary to make this test every time the alternators are to be paralleled; however, it is necessary to synchronize each and every time the alternator is to be paralleled; synchronizing connection is, therefore, made single phase.

C—Synchronizing test with lamps; the switch is left open, and the machine brought up to normal speed and frequency and the voltage is adjusted to the same as the bus voltage; when the speed and voltage have been adjusted to exactly equal the bus voltage and frequency, the lamps will be dark and the switch can be closed, connecting the alternator to the bus; if the speed should be slightly different, the lights will change alternately from light to dark; a condition may arise when the frequencies will be exactly the same, but the peaks of the voltage waves will be opposite, in which case the lights will remain light continuously; to correct this condition, it is necessary to change the speed of the incoming alternator until its voltage peak and the bus voltage peak occur at the same instant, at which point the lights will remain dark.

D—Three-wire 230—115 volt alternating-current distribution system.

E—Secondary network distribution; 208—120 volts; single unit.

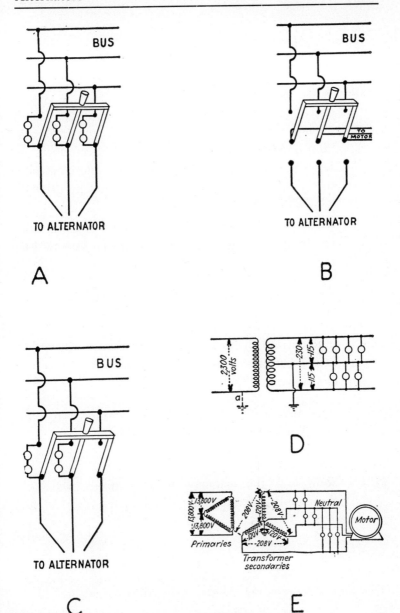

A

B

C

D

E

CLASS XI.   ELECTRICAL APPLIANCES

Section 63q.   Alternating-Current Motors

A—Variation of three-phase voltages covering two complete cycles; the numbers refer to the numbers in Figures B to N; each diagram shows the condition in the armature at the instant indicated by the corresponding number on this curve; the action of the magnetic field is smooth and regular; the rise and fall of currents in the conductors is also smooth and regular.

*B—The current entering the motor on line *1* divides equally and leaves the motor on lines *2* and *3*.

C—The current in line *2* is zero and that flowing in at line *1* leaves at line *3;* the magnetic field revolves clockwise.

D—The current in line *1* is small and, joining that from line *2*, flows out in line *3* which carries a maximum negative current.

E to N—Show how the magnetic field continues to rotate throughout the remainder of the cycle.

---

*Figures B to N show the electric and magnetic conditions in a two-pole, three-phase motor at the end of twelve equal parts of one cycle.

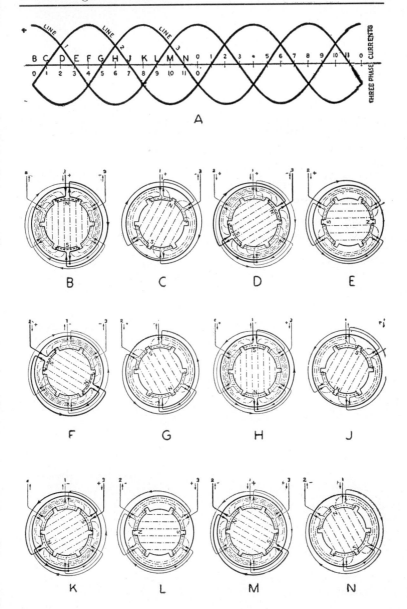

CLASS XI.   ELECTRICAL APPLIANCES

Section 63r.   Direct-Current Motors and Generators

A—Wiring diagram for a compound-wound motor or generator with interpoles which are in series with the armature so that their strength may be proportional to the load on the machine.

B—A series-wound generator in which all of the armature current passes through the field coils and through the external circuit; series motors are also connected in this manner.

C—Shunt-wound generator; part of the current generated in the armature is used exclusively to excite the field; this also applies to shunt-wound motors.

D—Compound-wound machine; combination of the shunt and series; it has two coils on each field coil, one series and one shunt coil.

E—Two compound-wound direct-current generators operating in parallel; in this way, they both contribute power to the same lines; it contains an equalizer, ammeters, and circuit breakers.

F—Edison three-wire system; the most common method of operating direct-current generators in series; either of two voltages is available with this system.

G—Direct-current motor starter, showing how the resistance units in series with the armature are cut out in starting and how the low-voltage magnet is connected to the field circuit of the machine.

H—Speed control of a direct-current motor by means of a resistance in series with the armature; more resistance gives lower speed.

J—Speed control of a direct-current motor by means of a resistance in series with the field; more resistance gives higher speed.

K—Starter and speed regulator.

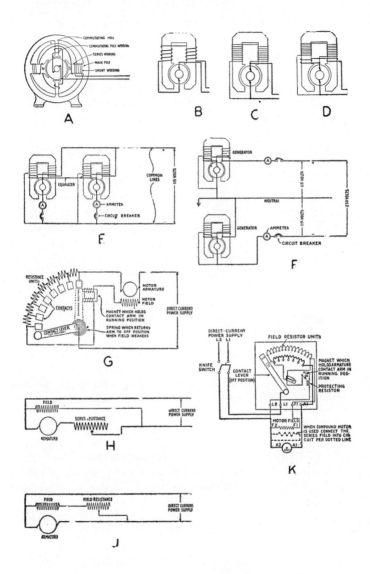

CLASS XI. ELECTRICAL APPLIANCES

Section 64a. Luminaires and Light Distribution

A, B, C—Shallow-bowl reflectors and shields; distribution downward zero; upward 75 per cent.

D, E, F—Translucent bowls; distribution downward 15 per cent; upward 65 per cent.

G, H, J—White glass enclosing globe or projecting luminous element with cased opal panels; distribution downward 45 per cent; upward 35 per cent.

K, L, M—Parabolic polished metal reflectors; distribution downward 50 per cent.

N, O, P—Large-area diffusing panels; distribution downward 55 per cent; extended trough reflector cased opal glass cover, enameled metal reflector with diffusing cover plates; distribution downward 70 per cent.

Q, R, S, T—High-boy open reflectors; prismatic mirrored or polished glass; distribution downward 70 per cent.

U, V, W—Large area diffusing reflectors; distribution downward 79 per cent.

X—Fundamental fluorescent-lamp circuit; fluorescent lamps are tubular bulbs coated inside by a phosphor powder, which is made to glow vividly by passing an electric current through mercury vapor inside the tube; the mercury-vapor emission is produced by electrodes at each end of the tube; fluorescent lamps require as auxiliary equipment a starting switch which momentarily opens and closes the electrode current and a ballast which consists of choke coils and condenses to limit the arc current and correct the power factor of the unit; fluorescent-lighting units generally include two or more tubes in order to provide more illumination and to reduce the stroboscopic effect.

Y—Nonuniform illumination resulting from wide-spaced units.

Z—Uniform distribution of illumination.

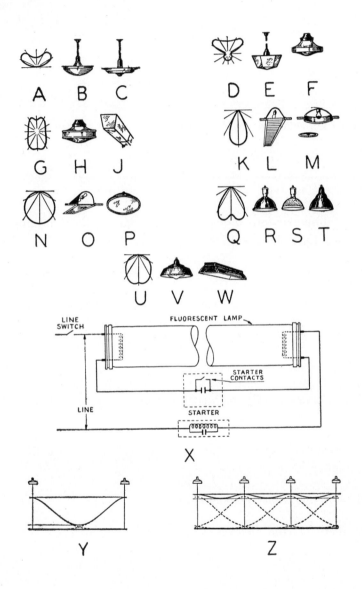

CLASS XI. ELECTRICAL APPLIANCES

Section 64b. Infrared Heating

A—Open reflector; gold plated or made of aluminum.

B—Closed reflector; vaporized aluminum.

C—Reflector-type lamp; vaporized aluminum.

D—Tunnel design; mounted over a slat conveyor for heating fairly large objects.

E—Tunnel design for heating a flat surface on a conveyor.

F—Tunnel design for heating small parts on a conveyor.

G—Method of heating a long suspended object.

H—Parabolic reflector emitting parallel rays.

J—Circular source at focus; reflector rays are redirected toward the center.

K—Elliptical reflector; the rays are redirected through the other focus.

L—Circular reflector; the source out of focus.

M—Paracyl (combination of parabolic and circular) reflector.

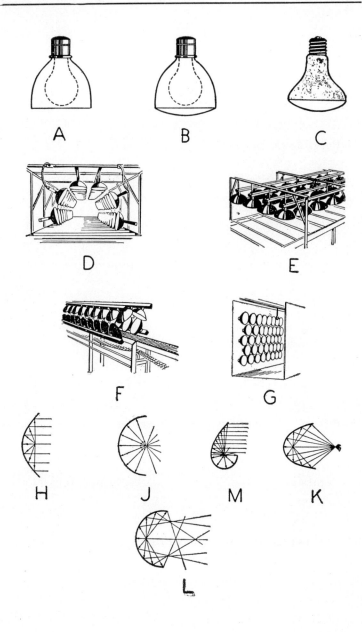

A

B

C

D

E

F

G

H

J

M

K

L

CLASS XI.   ELECTRICAL APPLIANCES

Section 64c.   Electric Heating

A—Portable radiant electric heater.

B—Radiant heater recessed in the wall.

C—Large industrial-type fan unit electric heater.

D—Large industrial-type portable fan unit electric heater.

E—Ceiling-mounted unit heater.

F—Wall-mounted unit heater.

G—Resistance-type boiler for steam or hot water.

H—Piping arrangement for connecting an electric water
    heater to a fire-box coil.

J—Domestic electrode-type hot-water heater for off-peak
   service.

K—Wiring diagram for a unit heater.

L—Arrangement of an electrode boiler.

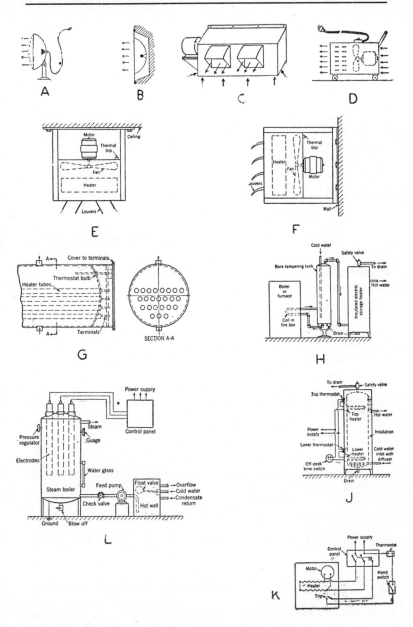

CLASS XI.   ELECTRICAL APPLIANCES

Section 65a.   Electromagnetic Apparatus

A— Hertzian oscillator; it consists of two metal plates and a rod connecting them interrupted by a spark gap; a spark traversing the gap makes it a good conductor, so that the vibrator may be considered as a single piece of metal along which the charge oscillates.

B— Electromagnetic waves just leaving a Hertzian oscillator.

C— Receiving apparatus; when electromagnetic waves impinge on a conductor, such as the aerial M, they set up oscillations in that conductor which may be detected by various means; the simplest of these is the crystal detector Z; T is a telephone or telegraph.

D— Crystal rectifying effect.

E— Deflection of a cathode ray by means of a magnet revealed by the effect on a fluorescent screen.

F— Deflection of cathode-ray tube; it is a long cylindrical vacuum tube containing two electrodes, a mica diaphragm with a slit at its center, and a long rectangular sheet of mica covered with a fluorescent salt; a discharge from the cathode passes through the slit and causes a stream of fluorescent light along the mica slit; that this stream consists of negatively charged particles can be shown by means of a strong magnet whose north pole will attract the stream and whose south pole will repel the stream (W. M. Welch Scientific Co.).

G— Focus tube; it consists of a vacuum tube with a concave cathode and a sheet of thin platinum foil at the center of the curved surface; when a discharge is sent through the tube, negatively charged particles fly off the cathode normally to the curved surface which centers, or "focuses" them on the platinum foil with sufficient energy to heat it red hot.

H— Crooke's tube; it consists of a pear-shaped vacuum bulb, 20 centimeters long and 7 centimeters in diameter, mounted horizontally on a wood base; the cathode is at the narrow end, the anode near the broad end and attached to it is a Maltese cross of metal arranged so that it may be set up in a vertical position; when in that position and a discharge is passed through the tube, the cross throws a sharp shadow on the glass, showing that the charged particles which make the glass fluoresce are stopped by the metal cross.

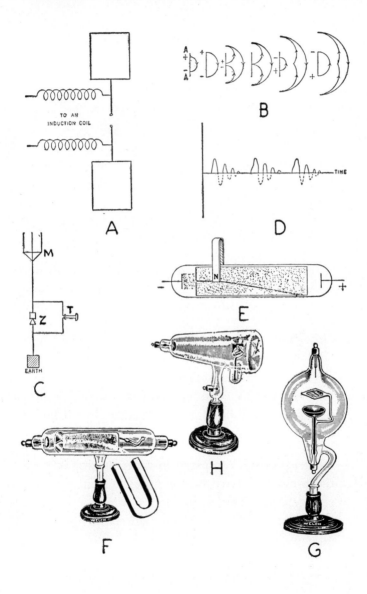

TO AN
INDUCTION COIL

A

B

TIME

D

M

T
Z

EARTH

C

N

−　　　　＋

E

WELCH

F

H

WELCH

G

CLASS XI.   ELECTRICAL APPLIANCES

Section 65b.   Wave and Electron Theory

A—Radio waves travel as electromagnetic and electro-static fields with the energy equally divided between the two; the electromagnetic lines of force and the electrostatic lines of force are at right angles to each other in a plane perpendicular to the direction of propagation; when a radio receiver picks up a radio wave, the wave induces a signal into the receiver at a value of possibly one millionth of a watt.

B—Cross section of a piece of copper wire though which no electric current is flowing; electrons are assumed to be moving in all directions.

C—Orderly parade of electrons caused by connecting the ends of the wire to the terminals of a battery; the electrons are made to drift from atom to atom toward the wire connected to the positive battery terminal; this orderly parade of electrons is an electric current; electrons are considered to be negative electricity.

D—Electric current compared to water flowing in a pipe; the pressure exerted on the water is analagous to the voltage; the quantity of water flowing is comparable to the current; the friction between water and the pipe is similar to the resistance.

E—Resistors in series.

F, G—Resistors in parallel.

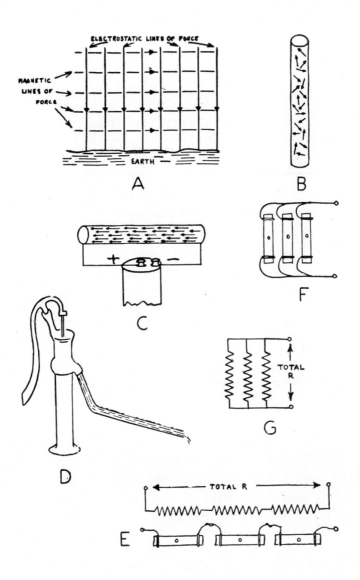

CLASS XI.   ELECTRICAL APPLIANCES

Section 65c.   Inductance and Capacitance

A—Electron action in a condenser; when voltage is applied, the electrons may be considered to be stretched out of place as shown by the dotted line; when the voltage is removed, the electrons return to their neutral position, causing a flow of current in opposite direction to the charging current.

B—Current flow through a condenser.

C—Current flow through an inductance.

D—Condensers in parallel.

E—Inductance and capacitance compared to a water system; when a surge of water from the pump strikes diaphragms Z, all of it cannot get through the small opening at the rate it comes from the pump and part of it is held back; this water then backs up into the storage spaces 1, 2, 3 when the surge stops, the flow through the diaphragm is maintained by the emptying of the storage spaces; the more storage spaces the steadier the flow, despite the fact that the pumping action itself is in surges.

F—Sine waves representing one cycle of alternating current or voltage.

G—Current and voltage relation in a-c circuits; E represents voltage; I represents current.

H—A typical application of series-tuned circuits, showing the tuned grid circuit of a radio-frequency or intermediate-frequency circuit.

J—Parallel resonance; in many radio circuits, the inductance is connected in parallel with the capacitance across the source of voltage.

K—Section of an electrolytic condenser.

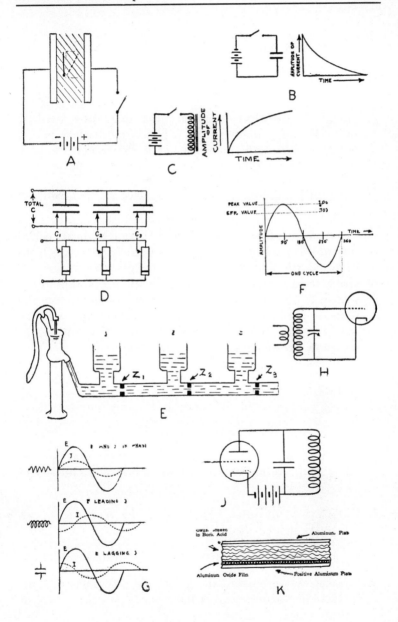

CLASS XI.   ELECTRICAL APPLIANCES

Section 65d.   Electron Tubes

A—Diode or two-electrode tube; the first tube used in radio, having a filament and a plate; one battery is connected across the filament for heating purposes; the positive pole of another battery is connected to the plate.

B—No positive voltage on the plate because the switch is not closed.

C—Electrons are driven out of the filament by heat; called thermionic emission; the filament which emits electrons is called the cathode; the body or plate to which the electrons are attracted is called the anode; as soon as the switch is closed and a small positive potential on the plate is applied, a steady stream of electrons will be attracted from the filament to the plate, causing a current flow around the circuit.

D—A stronger potential causes more electrons to flow until the plate is saturated. If the plate potential is reversed, no electrons will be attracted and no current will flow; if an alternating current is applied to the plate, the plate is alternately positive and negative; the plate current only flows while the plate is positive; therefore, current through the tube flows in one direction only and is said to be rectified.

E—Circuit for measuring the anode characteristic of an indirectly heated diode.

F—Connection of a diode with a directly heated cathode.

G—Typical anode current-anode voltage characteristics; all thermionic vacuum valves have anode characteristics of similar shape.

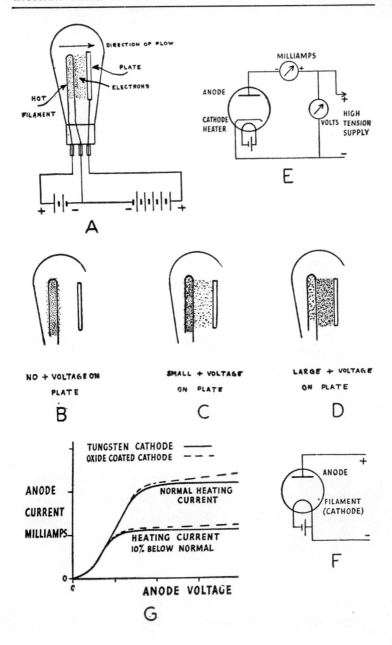

DIRECTION OF FLOW

PLATE

ELECTRONS

HOT
FILAMENT

A

MILLIAMPS

ANODE

CATHODE
HEATER

VOLTS

HIGH
TENSION
SUPPLY

E

NO + VOLTAGE ON
PLATE

B

SMALL + VOLTAGE
ON PLATE

C

LARGE + VOLTAGE
ON PLATE

D

TUNGSTEN CATHODE ———
OXIDE COATED CATHODE ----

ANODE
CURRENT
MILLIAMPS

NORMAL HEATING
CURRENT

HEATING CURRENT
10% BELOW NORMAL

0

ANODE VOLTAGE

G

ANODE

FILAMENT
(CATHODE)

F

CLASS XI.   ELECTRICAL APPLIANCES

Section 65e.   Hard Vacuum Tube Symbols

A—Filament L or heater and envelope.

B—Cathode K added.

C—Grid R added.*

D—Plate P added.

E—Screen and grid S G added.

F—Multigrid tube; the grids are numbered successively from the cathode toward the plate.

G—Heater-type diode; two diode plates DP_1 and DP_2 and a common cathode K.

H—Triode elements added to G form duodiode triode.

J—Base connections under view, indicating element connections to the base; the pins are numbered clockwise, starting at the locating key on octal tubes.

---

*Grids are shown either as broken or as zigzag lines.

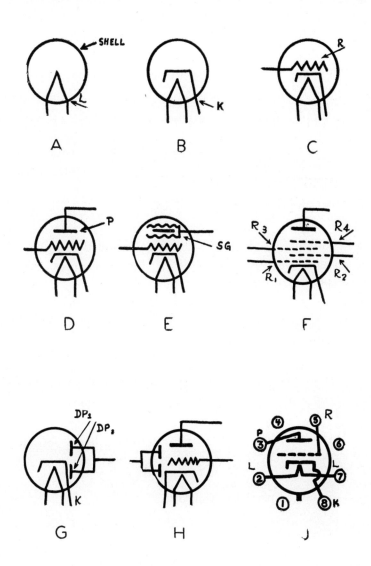

CLASS XI.   ELECTRICAL APPLIANCES

Section 65f.   Triodes, Screen-Grid Valves

A—Circuit for measuring triode characteristics.

B, C—Typical triode characteristics, the valve to which these curves refer has an anode resistance of 4,000 ohms, a mutual conductance of 4 milliamperes per volt and an amplification factor of 16.

D—Circuit of resistance-capacity coupled amplifying stage; the condenser "blocks" the direct-current component of the anode-to-cathode voltage.

E—Typical anode characteristics of a pentode or kinkless tetrode.

F—Typical characteristics of a screen-grid valve, showing the kink caused by secondary electron emission.

G—Construction of a kinkless tetrode (beam tetrode) dotted lines represent electron paths.

H—Enlarged partial section of G.

J, K—Details of a beam tube.

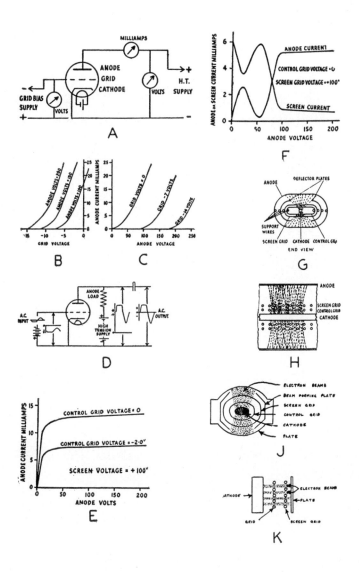

CLASS XI. ELECTRICAL APPLIANCES

Section 65g.   The Audio Amplifier

A—Transformer-coupled audioamplifier circuit.

B—Resistance-coupled audioamplifier circuit.

C—Single-power output audioamplifier circuit.

D—Push-pull output audioamplifier circuit.

E—Resistance-coupled phase inverter audioamplifier circuit.

F—Effects of audio-coupling condensers.

G—Effects of condensers and resistor, shunting audiocircuit.

H—Circuit for bass-tone compensation on volume control.

J—Circuit for treble-tone compensation on volume control.

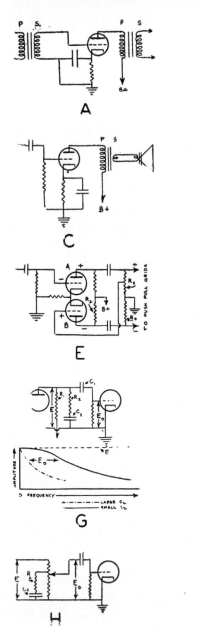

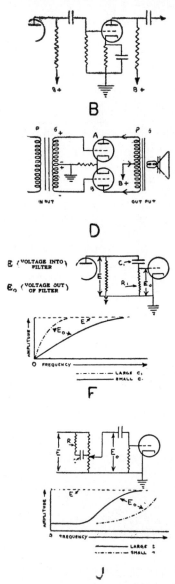

CLASS XI.   ELECTRICAL APPLIANCES

Section 65h.   Radio Communications System

A—Wave forms of an amplitude-modulated transmitter; simple diagram of the audiosignal and radio-frequency oscillator being combined in the modulator to form the transmitted signal.

B—Carrier which is modulated by a simple a-c voltage, starting at point X; actually this modulated wave would contain components of at least three radio frequencies; these would consist of the sum and difference of the original frequency and the modulated frequency.

C—Signal waves in a tuned radio-frequency receiver; variable tuned amplifier stages are adjusted by a single dial to the frequencies of different broadcasting stations between 540 kc to 1,600 kc.

D—Signal wave forms in a superheterodyne receiver; in a tuned radio-frequency receiver as in figure C, the receiver is tuned to the frequency of the signal and the signal is amplified at that frequency; while in a superheterodyne receiver the signal is tuned in and then changed in frequency to a lower value to which the intermediate amplifier is tuned and then the signal is amplified at the intermediate frequency.

E—Electrodynamic loud speaker; K indicates a cone; L, a voice coil; L_2, a field coil; S, a voice-coil support spider.

F—Permanent-magnet loud speaker; PM indicates a permanent magnet; S, a voice-coil support spider; K, a cone; L, a voice coil.

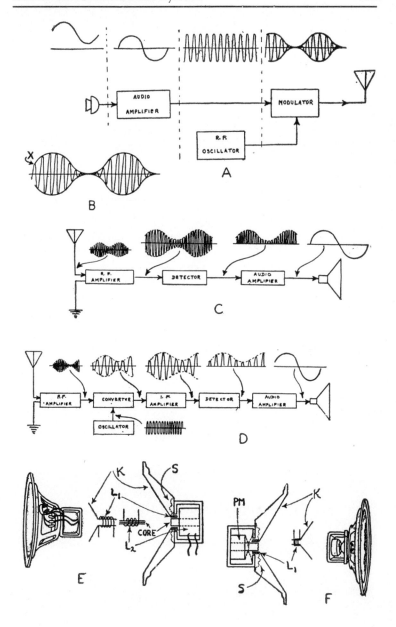

A

B

C

D

E

F

CLASS XI.   ELECTRICAL APPLIANCES

Section 65j.   Radio Communications Systems

A—Currents in a half-wave rectifier.

B—Currents in a full-wave rectifier.

C—Action of a filter.

D—An a-c d-c rectifier circuit.

E—Tuned radio-frequency amplifier circuit.

F—Circuit for neutralizing grid-to-plate capacity of a triode tube.

G—Circuit using a screen-grid tube.

H—Resistance-coupled radio-frequency amplifier circuit.

J—Tickler feedback oscillator circuit.

K—Oscillator circuit using single inductance.

L—Intermediate-frequency amplifier circuit.

M—Effect of a by-pass condenser across a cathode bias resistor.

N—Diode detector circuit.

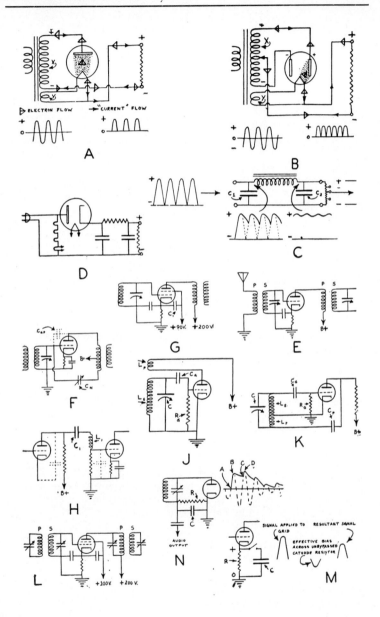

CLASS XI.   ELECTRICAL APPLIANCES

Section 65k.   Cathode and X-Ray Tubes

A—Electrostatic tube; electrons emitted from the cathode moving forward the final anode which is maintained at 500 to 5,000 volts, positive to cathode; the grid surrounding the cathode is made negative to it and repels the electrons some of which escape through a fine hole to form a "point source" of electrons. The number of electrons escaping and the brightness of the spot on the screen are controlled by the grid potential.

B, C—Electrostatic focusing with the electrostatic tube of figure A. C illustrates electric fields between the anodes; the arrows show the direction in which an electron is urged (the reverse of the conventional field direction); B shows the paths followed by electrons from the cathode to the screen.

D—Magnetic tube; it uses magnetic fields to focus and deflect the magnetic flux; in a cathode-ray tube of this type, an electron moving at right angles to a magnetic field is acted on by a force perpendicular to its direction of motion and perpendicular to the direction of the field; if emitted in a uniform field, in a perpendicular direction to the field, the electron moves in a circle the radius of which is proportional to its speed; the time to describe this circle is independent of the velocity of the electron and inversely proportional to the strength of the magnetic field; no force acts on an electron moving in the direction of a magnetic field; if the electron moves at an angle to a magnetic field, its velocity may be resolved into two components; the component along the field is unaltered; the component at right angles to the field is converted into a circular motion and the resultant path is a helix.

E—Magnetic focusing showing paths followed by electrons from a point source to a screen; five electrons are illustrated leaving the point source in different directions with different radial velocities; double-headed arrows indicate the magnetic field.

F—Arrangement of electrodes, in an X-ray tube.

G—Electron microscope; early 1940 model: magnification more than 100,000 times.

H—Magnetic electron lens.

J—Electrostatic electron lens.

K—Bragg-type X-ray spectrometer.

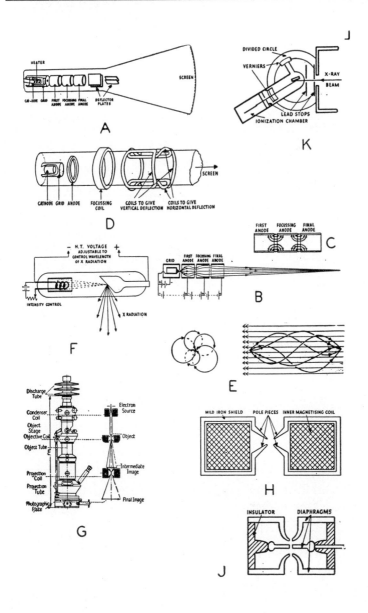

CLASS XI.   ELECTRICAL APPLIANCES

Section 651.   Gas-Filled Valves

A—Typical gas-filled tetrode; also called triode and marketed under various trade names, e.g., thyratron, gas-filled relay, or grid-controlled mercury-vapor relay; its properties differ from those of the hard vacuum valve and it has a different field of application.

B, C—Zigzag and spiral cathodes used in gas-filled valves to increase cathode voltage by conserving the radiated heat.

D—Gas-filled triode circuit and relay which operate when the light, which normally reaches the photocell, is cut off.

E—Operation of the relay in reverse direction to that of figure D; the relay operates when light reaches the photocell; used in the "Magic Eye" for opening and closing doors.

F—Speed control of a motor by variation of the phase angle between grid and anode voltage.

G—Speed control and reversal of the motor by means of two gas-filled triodes in a push-pull arrangement.

H—Synchronization between two rotors by means of a gas-filled triode.

J—Gas-filled triode used as an intermittent switching device or flasher.

K—Gas-filled triode used as a stroboscope.

L—Self-excited parallel converter; direct-current alternating-current input, output.

M—Series converter; direct-current alternating-current input, output.

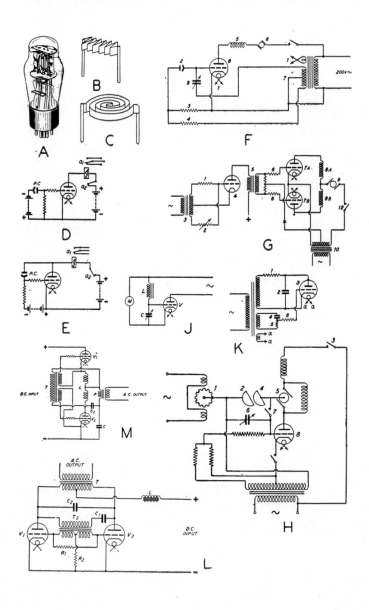

CLASS XI.   ELECTRICAL APPLIANCES

Section 66a.   Small Switch-Control Diagrams

A—Single pole, single break.

B—Single pole, double break.

C—Single pole, quadruple break.

D—Double pole, double break.

E—Simple two-way circuit giving on-off control at two separate points.

F—Two-way intermediate circuit giving on-off control at any number of points.

G—Two coupled two-way switches; equivalent to the intermediate circuit of figure F.

H, J, K—Conversion of an existing single-point on-off control to two points of control; A, existing circuit; B, correct method of conversion; C, incorrect method of conversion.

L, M, N—Variable control from two points, each switch comprising a pair of coupled three-way switches; B and C have no off position; if this is included, a two-way switch must be added at each point of control, and the circuit of figure O used; the two-way switches must be independent of the coupled three-way switches.

O—Variable control from three or more points; load A and B are shown selectively switched from three points with an off position at each point.

P—Knob switch giving the same result as two coupled three-way switches.

Q—Intermediate switch equivalent to three coupled three-way switch.

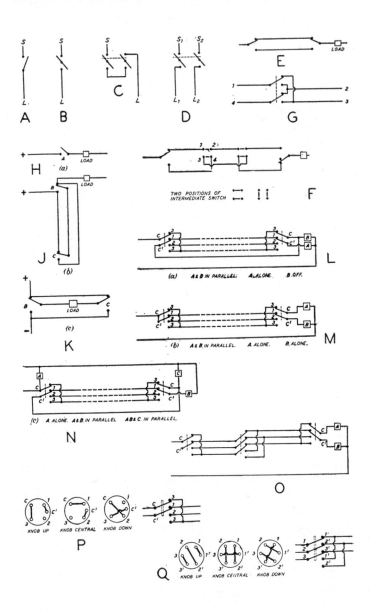

CLASS XI.   ELECTRICAL APPLIANCES

Section 66b.   Small Switch-Control Diagrams

A–Control for two loads at two points.

B, C–Mastering a two-way circuit either by A a single-way switch at a point distant from the normal points of control, or B at one of the existing points of control by the substitution of a three-way switch.

D–Restrictive master control.

E–Typical case of two-point control, with master on or master off when required; with both switches 1 and 2 open, the load is mastered off; with both closed, it is mastered on; with only one closed, a simple two-point on-and-off control is obtained at A or B.

F, G–Special cases of pilot circuits; A, pilot indication when any one of several circuits A, B and C is closed; B, pilot indication only when all these circuits are closed.

H–Equivalent of a series-parallel switch; two coupled two-way switches; loads 1 and 2 can be switched in series or parallel.

J–Series-parallel connection from two separate points (in off position).

K–Two resistance valves R_1 or $R_1 + R_2$ selected from either of two control points A or B.

L–Two independent two-way switches at one control point enabling R_1 or R_2 to be switched on alone in, series, or in parallel.

M–Load fed by battery A; battery B is a standby on charge from the mains; by using two coupled intermediate switches, the two batteries can be interchanged when A becomes discharged; this operation can be made automatic by a voltage-operated relay.

N–Two coupled two-way switches for reversing a small three-phase induction motor by interchanging two of the phases.

O–Face-plate stud switch; fifteen positions securing any combination of four independent circuits.

P–Drum-barrel switch for starting or reversing a series-wound motor.

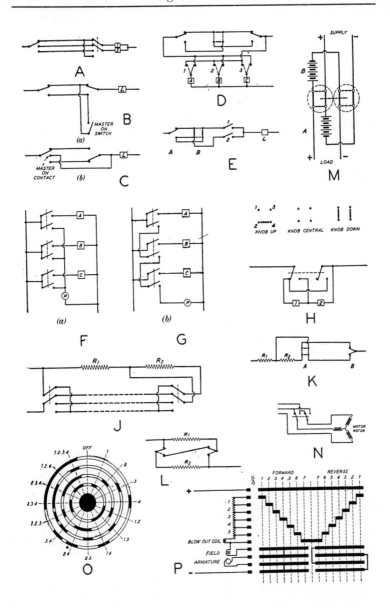

CLASS XI.   ELECTRICAL APPLIANCES

Section 66c.   Vacuum Switches

A—Operating a vacuum switch by means of an electromagnetic relay; excitation will open or close the switch as desired.

B—Manual operation of a vacuum switch with Bowden wire; suitable for remote control.

C—Operation of a vacuum switch by cam contact.

D—Burgess vacuum contact switch; a magnet *8* attracts an armature *10* attached to a lever *9* and pulls it down against the tension of a spring *11,* so that the switch contacts are in the position shown; when the current in coil *8* is cut off, the spring *11* pulls the lever *9* upward, and its pressure on a sleeve *2* opens the contacts *3* and *5;* the reverse operation may be obtained by arranging the armature on the opposite side of the sleeve *2,* so that excitation of the coil *8* will cause pressure of the arm *9* on the sleeve to open the contacts *3* and *5;* although the spark at the opening of contacts is very small, the heat may be sufficient to cause a slight volatilization, and the spark may not be visible in this type of switch; the pressure necessary for positive operation is small, so that it will work easily with the mechanical movement of a standard telephone relay; the path of the extremity of the stem is about 0.02 inch long and the elasticity of the glass permits repeated movements many times a second.

E—Thermally operated vacuum switch (Sunvic, Great Britain).

F, G, H—Typical methods of connecting vacuum switches; *A* load contacts are closed by closing the control switch; *B* load contacts are closed by opening the control switch; *C* three-phase load controlled by two switches with their control windings on one phase or in an independent circuit.

J—Simple and flexible vacuum switch operated by a very light pressure; housing of contacts in a glass envelope under vacuum prevents arcing when a circuit is broken; arcing would cause pitting and corrosion.

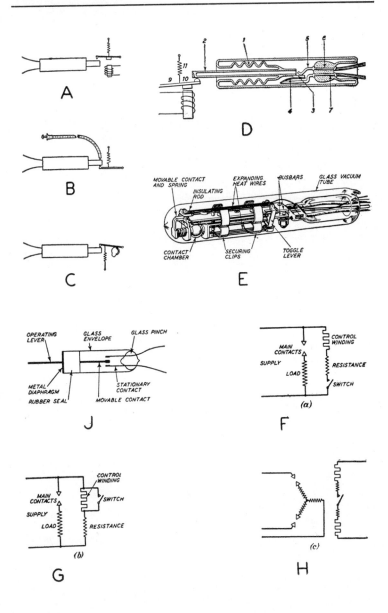

A

B

C

D

E — MOVABLE CONTACT AND SPRING · INSULATING ROD · EXPANDING HEAT WIRES · BUSBARS · GLASS VACUUM TUBE · CONTACT CHAMBER · SECURING CLIPS · TOGGLE LEVER

J — OPERATING LEVER · GLASS ENVELOPE · GLASS PINCH · METAL DIAPHRAGM · RUBBER SEAL · STATIONARY CONTACT · MOVABLE CONTACT

F — MAIN CONTACTS · SUPPLY · LOAD · CONTROL WINDING · RESISTANCE · SWITCH (a)

G — MAIN CONTACTS · SUPPLY · LOAD · CONTROL WINDING · SWITCH · RESISTANCE (b)

H — (c)

CLASS XI.   ELECTRICAL APPLIANCES

Section 66d.   Relays

A to E—Symbols for relay contacts; contacts are always shown, unless otherwise stated, in the position they take up when the supply is disconnected; A, makes contact completing circuits *1* and *2;* B, break contact, isolating circuits *1* and *2;* C, change-over contact, first isolating *1* and *2* and then completing circuits *1* and *3;* D, make before break, completing circuits *1, 2* and *3,* and then isolating *2;* E, break, make before break, first isolating circuits *1* and *2,* then completing circuits *1, 3* and *4,* finally isolating *3.*

F—Preventing contact rebound.

G—Stepped relays with multiple-contact bank; sequence *a* closes, *b* opens, *c* and *d* change over.

H—Sequence of relays; *a* closes, *b* and *c* change over, *a* closes, *e* closes; this arrangement insures that the uppermost contacts close only after all the others have operated.

J—Domed contacts.

K—Flat contacts.

L—Siemens type 73 high-speed relay; it greatly reduces the time of making contact.

M—Slow-releasing relay with a heel slug at the end of the yoke which is remote from the armature; the operation of this relay is not affected when the coil circuit is made but a delayed release is secured when the circuit is opened.

N—Slow-operating relay with copper slug at the armature end of the coil; the broken arrow indicates the path of the magnetic flux due to the main coil at the instant of closing the coil circuit.

O—Polarized relay; the armature *2* makes contact either on the left or on the right, according to the direction of the current through the field coil.

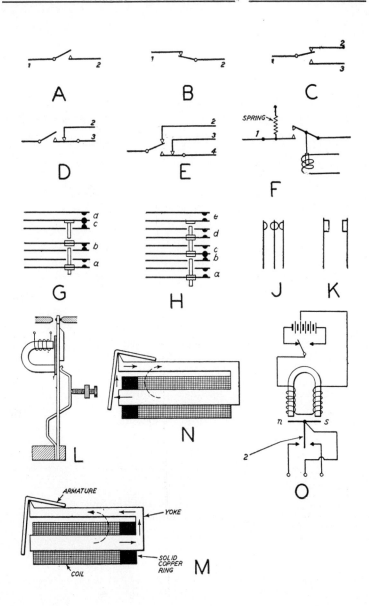

CLASS XI. ELECTRICAL APPLIANCES

Section 66e. Mercury Switches

A—Single make-and-break contact with fused liner.

B—Single make-and-break switch with change-over contacts.

C—Switch with top connection and hard-glass liner.

D—Same as C but with change-over contacts.

E, F—Vertical-type, coil-operated contacts.

G, H—Reverse-acting, vertical-type, coil-operated contacts.

J—Mercury switch operated by an electromagnet.

K—Diagram illustrating the use of a three-contact mercury switch for breaking an inductive direct-current circuit.

L, M—Centrifugal switches designed for rotation about a vertical axis; L opens the contacts; M closes the contacts, when the speed of rotation exceeds a given rate, by forcing the mercury upward in the bulb.

N—Inclination switch; the normally open contacts are completed if the vertical axis of the tube is inclined about 30° in any vertical plane.

O—Magnetically operated nontilting switch; a light iron rod bridges the mercury pools; an external electromagnet raises this rod to open the circuit.

P—Acceleration switch; the contacts will close if the moving object to which the switch is attached is accelerated in the direction of arrow 1 or retarded in the reverse direction.

Q—Switch for indicating rotation in one direction; the contacts are closed once during every revolution in the direction shown, but not in the reverse direction.

R—Switch for indicating movement in a certain direction; tilting the switch toward the right closes the contacts, which remain closed until the switch is subsequently tilted in the opposite direction.

S, T, U, V, W—Various forms of mercury switches with delayed action.

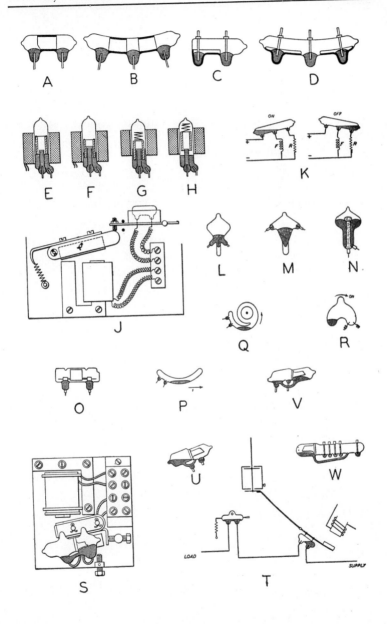

CLASS XI.   ELECTRICAL APPLIANCES

Section 67a.   Radar Navigational Equipment

A—Radar circuits; the radar antenna rotates continuously and indicates any obstruction in its path; this indication appears as a spot on a cathode-ray oscillator, known as a "plan position indicator" but more often designated as the "scope" which appears as a bright spot as shown at A in figure B.

B—The "scope"; the position of the spot with reference to the compass card in the outside rim of the scope face tells the direction of the obstruction, while the distance of the spot from the center of the scope indicates the distance of the object from the point of observation; the basic principle of operation is that of reflection of radio waves generated by the magnetron.

C—Ship passing through marked by two buoys.

D, E, F—Three different navigational conditions indicated on the scope: figure D shows the condition of steering so that the ship will be heading directly for the middle of the opening figure; E shows that the right-hand echo spot, from the starboard buoy, is farther away from the scope O-line than the echo spot from the port buoy; figure F shows the reverse.

G—Sailing between two cliffs.

H—Scope presentation of figure G.

J—Sailing nearer to one side of the cliff.

K—Scope presentation of figure J.

L—Ship passing cliff walls and channel which are not straight.

M—Scope presentation of figure L.

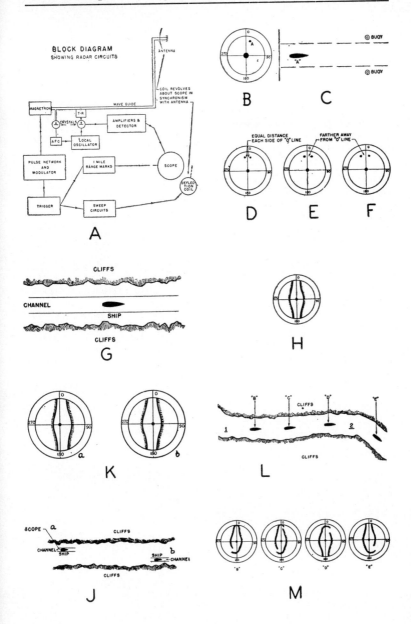

CLASS XI.   ELECTRICAL APPLIANCES

Section 67b.   Echo Sounding

A—Echo sounding recorder used for depth measurement; a short pulse of sound is sent out from the bottom of a ship and the time is measured for the resulting echo to return from the bottom of the sea; in water, sound waves have a propagation speed of about 4,855 feet per second, which is nearly four times faster than the sound propagation in air; the illustration shows a typical installation schematically: B, gear train; C, switch cam; D, stylus; E, front of tank; F, automatic governor; G, transmitting contact; H, motor.

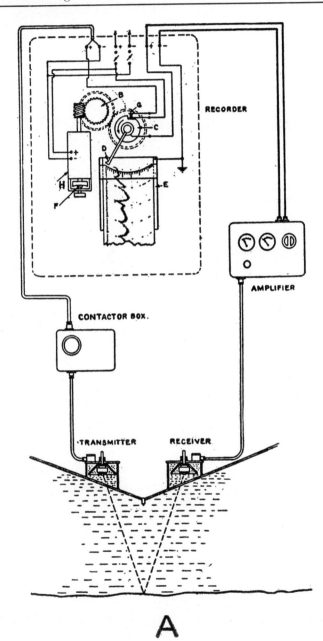

RECORDER

B

G

C

D

E

H

F

AMPLIFIER

CONTACTOR BOX.

TRANSMITTER   RECEIVER

A

CLASS XII.   COMFORT HEATING, COOLING, AND AIR CONDITIONING

Section 68a.   Steam-Heating Hookups

A—Vapor heating system; two-pipe up-feed shown; the radiators discharge condensate to the dry return pipe; these systems operate at a few ounces pressure and above, but those with mechanical return devices may operate at pressures up to 10 psig; control of heat is obtained by varying the opening of the graduated radiator valve; the boiler is maintained at a constant pressure slightly above atmospheric.

B—Vacuum heating system; a vacuum is maintained in the return line at all times; the pump is usually controlled by a vacuum regulator which operates the pump to maintain the vacuum within limits in response to a pressure difference between the atmosphere and the vacuum in the return main.

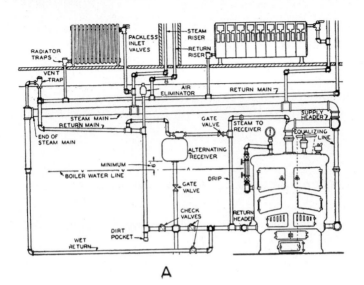

A

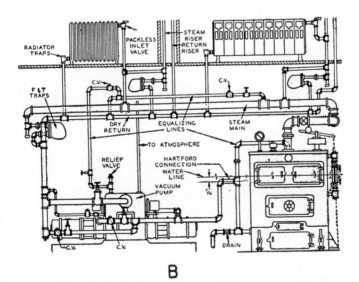

B

CLASS XII.   COMFORT HEATING, COOLING, AND AIR CONDITIONING

Section 68b.   Steam-Heating Hookups

A—Arrangement of a strainer ahead of the trap.

B—Arrangement of a dirt pocket ahead of the trap.

C—Dripping down-feed riser; the trap discharges directly into the return line.

D—Vent and drip on the end of the main for a wet return system.

E—Maintaining a wet drip line and draining through a float trap to a vacuum pump or waste.

F—Sarco thermostatic steam trap No. 9-125 for pressures of 0-125 psig; *1*, body; *2*, cap; *3*, cap screws; *4*, cap gasket; *5*, lock washer; *6*, element; *7*, valve head; *8*, valve seat; *9*, seat gasket.

G—Radiator connections where the run-outs are on the ceiling of the floor below.

H—Radiator connections where the run-outs are in the floor construction.

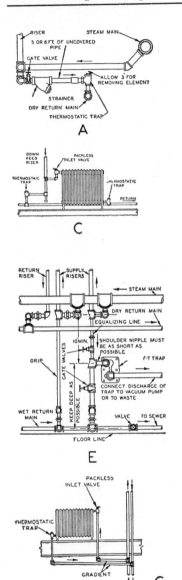

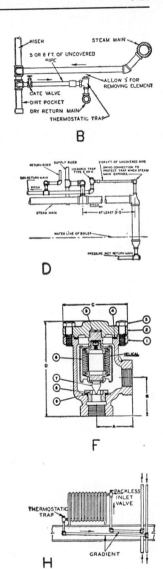

CLASS XII.   COMFORT HEATING, COOLING, AND AIR CONDITIONING

Section 68c.   Steam-Heating Hookups

A—Looping the dry return at doorways.

B—Connections for basement radiators below the dry-return main.

C—Hartford connection and recommended boiler connection for single low-pressure heating boiler.

D—Alternating receiver lifting the condensate from the heat exchanger.

E—Float thermostatic trap dripping base of a steam riser.

F—Typical steam run-out when the risers are not dripped.

G—Typical steam run-out when the risers are dripped.

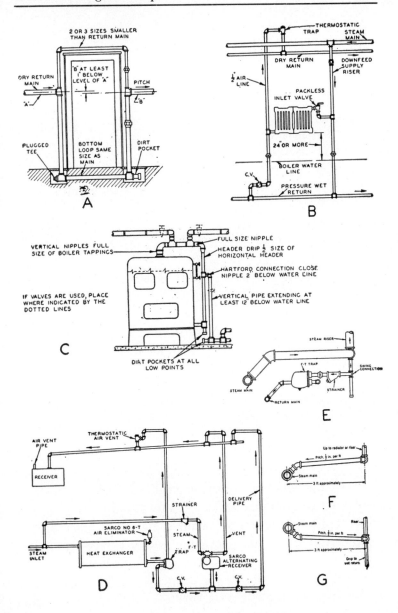

CLASS XII.   COMFORT HEATING, COOLING, AND AIR CONDITIONING

Section 68d.   Steam-Heating Hookups

A—Sarco liquid-level controller, maintaining the water level in a steam boiler.

B—End of main dripped through a float thermostatic trap.

C—Blast surface equipped with float thermostatic traps and strainers.

D—Draining a heating-pipe coil.

E—Dripping floor type low-pressure unit heater and supply line.

F—Bucket-trap dripping end of a steam main.

G—Making lifts on vacuum systems when the distance i• over 5 feet.

H—Detail of the main return lift at the vacuum pump.

J—Changing the size of a steam main when the run-outs are taken from the top.

K—Discharge of high-pressure apparatus into low-pressure heating mains and vacuum return mains through a low-pressure trap.

L, M—Acceptable and preferred methods of branching from the main.

N—Looping the main around a beam.

O—Dripping the end of the main into a wet return.

P—Dripping the main where it rises to a higher level.

Q—Reducing the size of a main at a swing connection.

R—Dripping the end of a main into a dry return; a gate valve is recommended at the inlet side of the trap.

S—Reducing the size of the main.

T—Dripping the heel of a riser into a dry return; a gate valve is recommended at the inlet side of the trap.

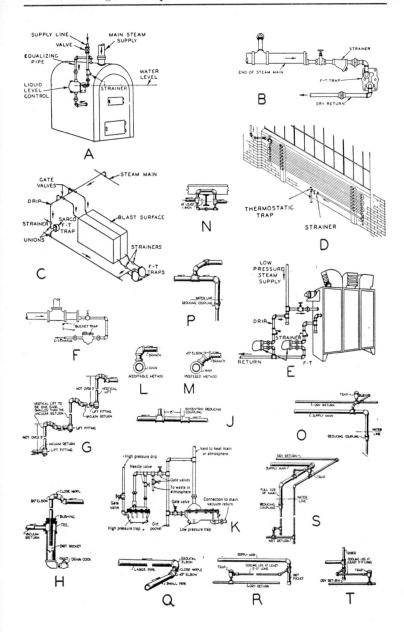

CLASS XII.   COMFORT HEATING, COOLING, AND AIR CONDITIONING

Section 69a.   Hot-Water Heating Systems

A—Direct-return two-pipe hot-water system.

B—Reversed-return two-pipe hot-water system.

C—Open expansion tank.

D—Closed expansion tank.

E—Gravity hot-water system; two-pipe circulation.

F—Forced-circulation hot-water system; two-pipe direct return.

G—Hot-water supply with overhead distribution to the radiators.

H—Down-feed hot-water supply to the top of radiators with bottom return.

J—Up-feed hot-water supply to the top of radiators with bottom return.

K—One-pipe gravity-circulation hot-water system.

L—One-pipe forced-circulation hot-water system.

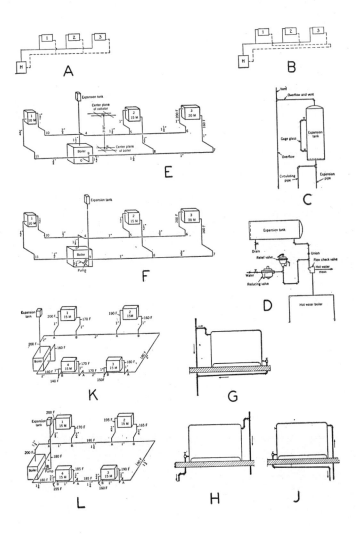

CLASS XII.   COMFORT HEATING, COOLING, AND AIR CONDITIONING

Section 69b.   Radiant Hot-Water Heating

A—Hot-water temperature control of a radiant heating surface by Johnson Service Co.; Duostat; an indoor-outdoor thermostat and a high-limit thermostat control the steam supply to the hot-water heater.

B—Controlling a three-way mixing valve on a hot-water supply to pipe coils in a radiant heating surface with room thermostat and high-limit thermostat controls; automatic firing unit on a hot-water boiler.

C—Zone control of radiant heating surfaces; an indoor-outdoor thermostat and a high-limit thermostat control a diaphragm valve in the steam supply to the hot-water heater; room thermostats control the diaphragm valves in the water supply to the individual room coils.

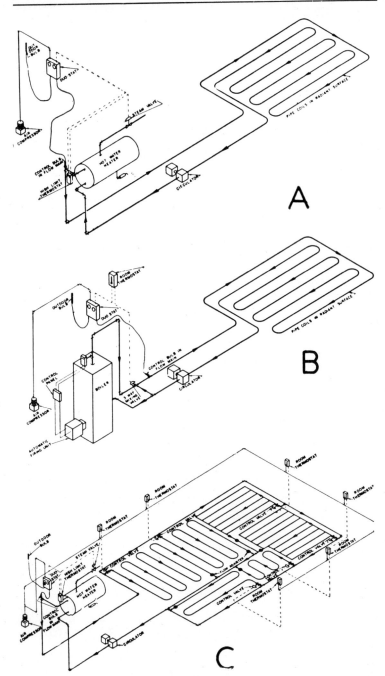

CLASS XII.   COMFORT HEATING, COOLING, AND AIR CONDITIONING

Section 69c.   Radiant Heating*

A—Radiant heating coil laid on a lath which is attached to the joists in the usual manner; rock wool or any other loose insulating material is then laid over the pipe between the joists.

B—Heating coil hung below the joists with the metal-mesh lathing wired to the pipe; plaster is then applied to the lath and covers the pipe as indicated.

C—Radiant heating coil embedded in a concrete floor panel over reinforcing supported by a steel beam or joist.

D—Heating coil laid on top of joists with wood flooring supported by sleepers laid over the coil; some type of insulating board is used between the joists as shown; this board is supported by stops nailed to the joists.

E—Radiant heating coil embedded in concrete which has been poured on a bed of packed gravel.

F—Heating coil in packed gravel over which the concrete floor has been poured.

G—Heating coil laid on top of concrete and wood flooring supported by sleepers laid over the coil.

H—Sand is leveled on a bed of packed gravel: radiant heating coil laid on the sand and the concrete floor poured.

J—Square coil; hot lines are on the exterior of the room; it is cooler in the center.

K—Grid coil; minimum bending is required.

L—Continuous coil in parallel with low friction-head loss.

M—Continuous coil; minimum welding is required.

*Reprinted by permission of National Tube Co.

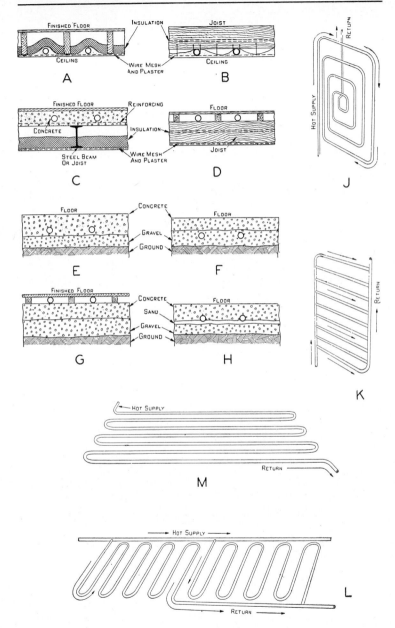

CLASS XII.   COMFORT HEATING, COOLING, AND AIR CONDITIONING

Section 70a.   Warm-Air Heating System*

A—Gravity warm-air heating system; when an air-circulating fan is inserted in the duct supply system it is called a forced-air system; *A,* the house chimney, no bends nor offsets; *B,* top of chimney at least 2 feet above the ridge of the roof; *C,* flue lining, fireclay; *D,* all joints air tight; *E,* at least 8 inches brick; *F,* no other connection beside that to the furnace; *G,* clean-out frame and door, air-tight; *H,* smoke pipe, end flush with inner surface of flue; *I,* draft door; *J,* flue thimble; *K,* casing body; *L,* casing hood or bonnet, the top of all leader collars on the same level; *M,* round leader, pitch 1 inch per foot; *N,* sleeve with air space around the leader where passing through the wall; *O,* dampers in all leaders; *P,* transition fittings; *Q,* rectangular wall stack; *R,* baseboard register; *S,* pipes distributed equally around the bonnet; *T,* floor register; *U,* return air face; *V,* panning under joist; *W,* transition collar; *X,* round return pipe; *Y,* transition shoe; *Z,* the top of the shoe at the casing not above grate level.

*Reprinted by permission from N.W.A.H. & A.C. Assn. Standard Code.

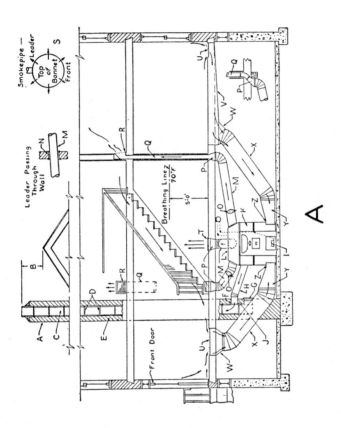

Smokepipe — Leader

S

Top of Bonnet

Front

Leader Passing Through Wall

N
M

B

A
C

D

E

Front Door

R
Q

R

Q

P

Breathing Line 70°F

5'-0"

U

W

P
Q

X

M

Z

O
K
Y

T
P

M
O
F
H
G
Z
Y
J
X

U

W

A

CLASS XII.   COMFORT HEATING, COOLING, AND AIR CONDITIONING

Section 71a.   Unit Heaters and Ventilators

A, B, C—Centrifugal-fan floor-mounted unit heaters.

D—Centrifugal-fan suspended unit heater.

E—Propeller-fan unit heater with horizontal blow.

F—Propeller-fan unit heater with downward vertical blow.

G—Unit-heater connection to a one-pipe gravity steam system.

H—Unit-heater connection to a gravity steam system with wet and dry return.

J—Connection of a unit heater to a high-pressure steam system.

K—Connection of a unit heater to a hot-water system.

L—Draw-through unit ventilator and heater.

M—Blow-through unit ventilator and heater.

N—Typical window ventilator without a heating coil.

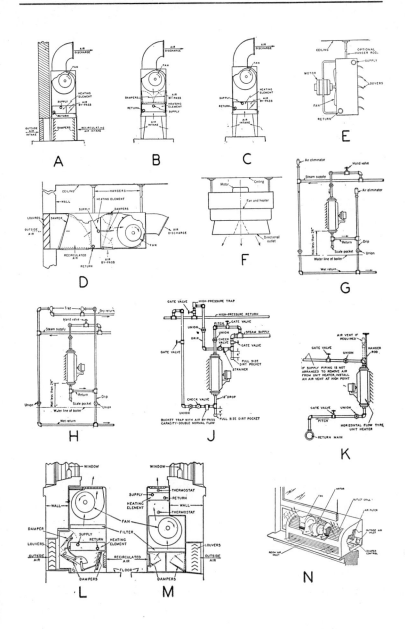

CLASS XII   COMFORT HEATING, COOLING, AND AIR CONDITIONING

Section 72a.   Unit Air Conditioners

A—Spray-type unit humidifier with steam coil to preheat the air for residences.

B—Spray-type remote all-year-round air conditioner.

C—Self-contained all-year-round air conditioner with water-cooled refrigerating condenser and compressor.

D—Self-contained summer air conditioner with air-cooled refrigerating condenser and compressor.

E—Remote floor-type room-unit air conditioner.

F—Vertical remote unit air conditioner.

G—Suspended propeller-fan room cooler.

H—Humidifying unit for a warm-air furnace.

J—Surface-type cooling unit.

K—Brine-spray cooling unit.

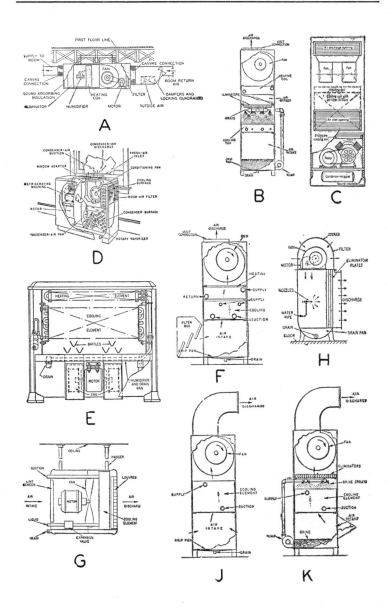

CLASS XII.   COMFORT HEATING, COOLING, AND AIR CONDITIONING

Section 72b.   Unit Air Conditioners

A—Humidifying unit for steam-radiator-heated homes.

B—Oil-fired air-conditioning unit.

C—Air-conditioning unit with top inlet and outlet.

D—Residential, central Carrier year-round air-conditioning unit equipped for gas-burning with steam boiler, refrigerating coil, and heating coil.

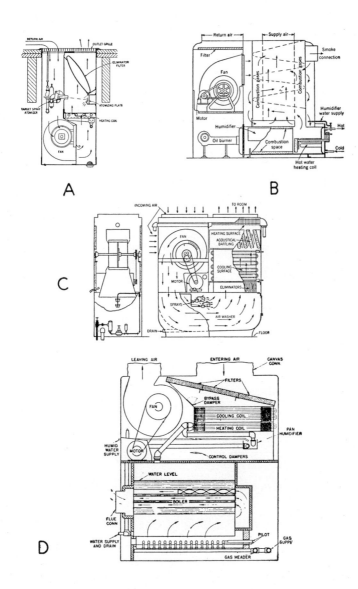

CLASS XII. COMFORT HEATING, COOLING, AND AIR CONDITIONING

Section 73a. Temperature Controls

A—Diaphragm thermostat.

B—Direct-expansion thermostat.

C—Straight-strip bimetallic thermostat.

D—Spiral bimetallic thermostat.

E—Curved-strip bimetallic thermostat.

F—Mercury-tube switch.

G—Expanding bellows.

H—Cooling switch which closes one circuit on temperature rise.

J—Heating switch that closes one circuit on temperature fall.

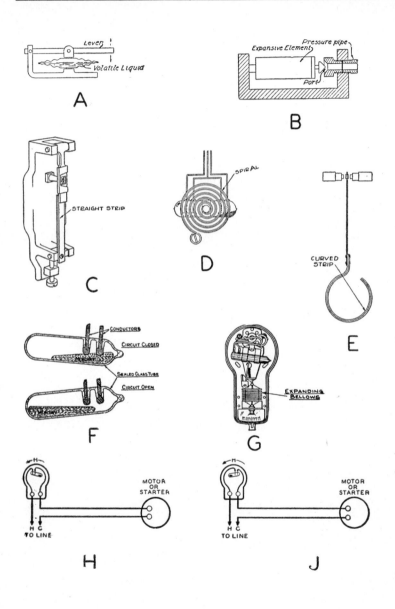

CLASS XII.   COMFORT HEATING, COOLING, AND AIR CONDITIONING

Section 74a.   Refrigeration Systems

A—Direct system.

B—Indirect open-spray system.

C—Indirect closed-surface system.

D—Indirect, vented closed-surface system.

E—Double, indirect, vented open-spray system.

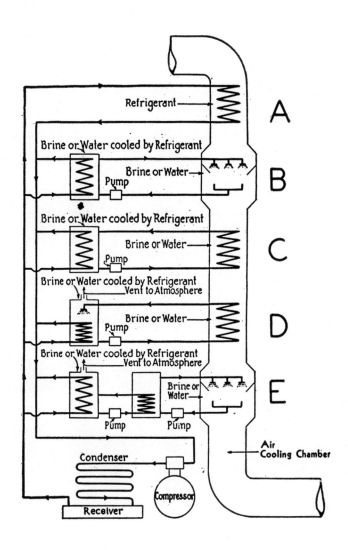

CLASS XII.   COMFORT HEATING, COOLING, AND AIR CONDITIONING

Section 74b.   Compression System of Refrigeration

A—Air conditioning; with compression refrigeration, when mechanical means are used, the system may be an ordinary direct expansion system, a steam-jet system or a water-vapor system; the most common type is that employing direct expansion, where a power-driven piston-type compressor is used to compress the gas, a condenser to liquefy the gas, and a cooling coil in which the gas expands after entering the coil through an orifice, e.g., an expansion valve; various other accessories are added, such as a receiver, an oil separator, a liquid trap, etc.; Freon is most commonly used in such systems as a refrigerant because of its comparative harmlessness, although ammonia is a more efficient refrigerating medium.

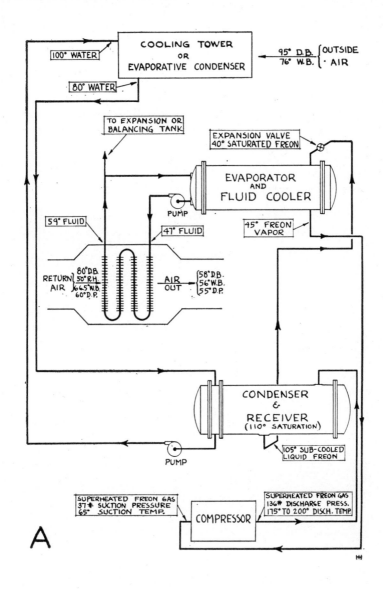

CLASS XII.   COMFORT HEATING, COOLING, AND AIR CONDITIONING

Section 74c.   Reverse-Return Chilled-Water System

A—Reverse-return water-cooling system; a three-pipe re-frigerated water-circulating system which delivers chilled water to remote unit air conditioners at a constant de-livery pressure; water may be heated in a separate tank and circulated as in a hot-water heating system for heating during the cold seasons.

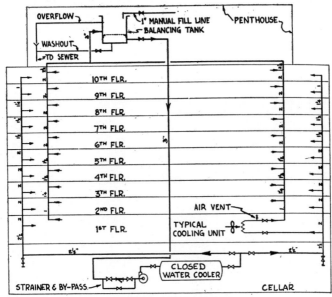

RISER DIAGRAM

Note: Allow 4 GPM for each cooling unit. (Equivalent to 2 tons refrigeration)

TABLE : PIPE SIZING OF WATER MAINS AND BRANCHES (Based on 2 GPM per ton of refrigeration and 12° rise to and from water cooler.)				
NOMINAL PIPE SIZE	GPM	HEAD LOSS FT. PER 100	VELOCITY FT. PER SEC.	TONS REFRIGERATION
10	2400	5.04	9.76	1200
8	1200	4.90	8.33	600
6	750	4.46	6.66	375
5	360	4.24	5.77	180
4	220	5.12	5.55	110
3½	150	4.87	4.67	75
3	100	4.47	4.34	50
2½	55	4.24	3.69	22.5
2	36	4.37	3.35	18
1½	18	4.31	2.84	9
1¼	10	3.08	2.15	5
1	5	3.24	1.86	2.5

CLASS XII.   COMFORT HEATING, COOLING, AND AIR CONDITIONING

Section 74d.   Typical By-Pass Systems

A—Single-pass system; all of the conditioned air will pass through the cooling coil or air washer.

B—Face and by-pass system; dampers are usually placed across the surface of the cooling coil or washer chamber, and a by-pass duct with damper permits part of the air to pass around the cooling coil or washer; these dampers are generally operated together, so that as one closes the other opens.

C—Return-air by-pass; instead of the by-pass handling a mixture of outside and return air, the by-passed air comes directly from the return-air duct and dampers are installed at that point; the air that is by-passed and mixed with conditioned air is return air from the rooms being served.

D—Outside air by-pass.

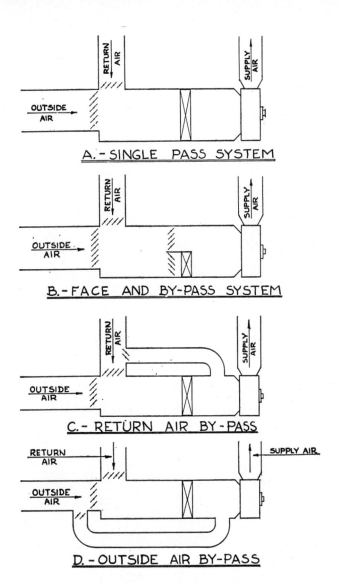

A. – SINGLE PASS SYSTEM

B. – FACE AND BY-PASS SYSTEM

C. – RETURN AIR BY-PASS

D. – OUTSIDE AIR BY-PASS

CLASS XII.   COMFORT HEATING, COOLING, AND AIR CONDITIONING

Section 74e.   Sheet-Metal Details

A, B, C—Fan-discharge connections.

D, E, F—Filter, washer, cooler, and heater connections.

G—Easement around obstructions.

H, J—Branch take-offs.

K—Sheet metal sections.

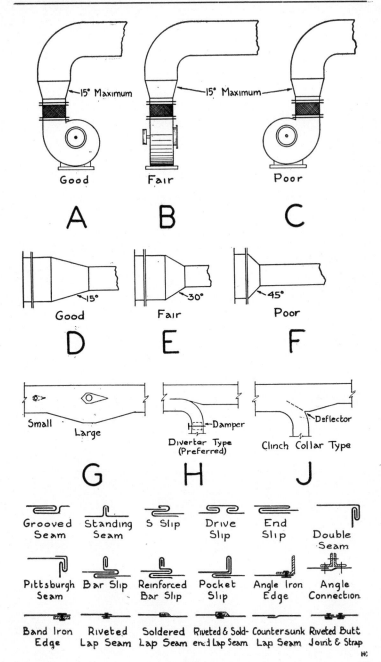

Good — A

Fair — B

Poor — C

15° Maximum

15° Maximum

Good — D — 15°

Fair — E — 30°

Poor — F — 45°

Small — Large — G

Damper
Diverter Type
(Preferred) — H

Deflector
Clinch Collar Type — J

Grooved Seam — Standing Seam — S Slip — Drive Slip — End Slip — Double Seam

Pittsburgh Seam — Bar Slip — Reinforced Bar Slip — Pocket Slip — Angle Iron Edge — Angle Connection

Band Iron Edge — Riveted Lap Seam — Soldered Lap Seam — Riveted & Soldered Lap Seam — Countersunk Lap Seam — Riveted Butt Joint & Strap

K

CLASS XII.   COMFORT HEATING, COOLING, AND AIR CONDITIONING

Section 74f.   Humidifiers

A—Steam grid humidifier.

B—Enclosed steam grid humidifier.

C—Steam pan humidifier.

D—Electric pan humidifier (Johnson Service Co.).

E—Plan of an air washer, circulating pump and spray chamber; when supplied with preheated water or air, it acts as a humidifier and increases the absolute humidity of the air stream.

F—Side elevation of the apparatus in figure E.

G—End elevation of the apparatus in figure E; eliminators and motor are not shown.

H—Eliminator details.

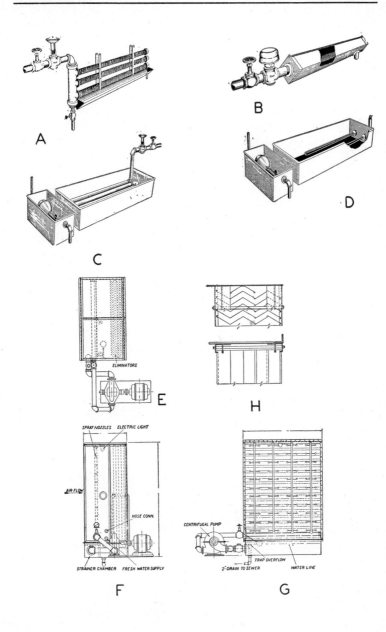

A

B

C

D

ELIMINATORS

E

H

SPRAY NOZZLES   ELECTRIC LIGHT

AIR FLOW

HOSE CONN.

STRAINER CHAMBER   FRESH WATER SUPPLY

F

CENTRIFUGAL PUMP

TRAP OVERFLOW

2"DRAIN TO SEWER   WATER LINE

G

CLASS XII.   COMFORT HEATING, COOLING, AND AIR CONDITIONING

Section 75a.   Heat-Pump Systems

A—Air-to-air heat-pump system hooked up for cooling effect; valves on dotted lines closed, thus permitting operation as a refrigeration system in which heat at a temperature level too high for use as a cooling effect is *extracted* in the evaporator (inside surface), pumped by the refrigeration mechanism to a higher temperature and rejected by the condenser (outside surface) to a condensing medium usually water, but, in this case, the outside air. The heat-pump system is frequently referred to incorrectly as the reverse cycle system.

B—Air-to-air heat-pump system hooked up for heating effect; valves on dot-dash lines closed thus permitting operation as a heat-pump system in which heat at a temperature level too low for heating purposes is *absorbed* in the evaporator (outside surface), pumped by the refrigeration mechanism to a higher temperature and rejected by the condenser (inside surface) to a condensing medium usually water, but, in this case, the desired warm air.

C—Water-to-water heat-pump system hooked up for cooling effect in which the cooled air desired is passed over pipe coils through which cooled water is recirculated. The condenser water (c.w.) from an outside source is rejected to the drain.

D—Water-to-water heat-pump system hooked up for heating effect in which warm air desired is passed over pipe coils through which warm condenser water is recirculated. The water from the heat source passes through the water cooler and is rejected to the drain.

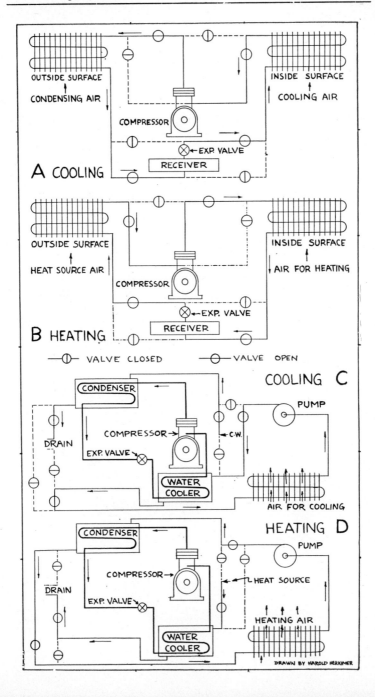

OUTSIDE SURFACE

CONDENSING AIR

COMPRESSOR

EXP. VALVE

RECEIVER

INSIDE SURFACE

COOLING AIR

A COOLING

OUTSIDE SURFACE

HEAT SOURCE AIR

COMPRESSOR

EXP. VALVE

RECEIVER

INSIDE SURFACE

AIR FOR HEATING

B HEATING

VALVE CLOSED   VALVE OPEN

COOLING **C**

CONDENSER

COMPRESSOR

PUMP

C.W.

DRAIN

EXP. VALVE

WATER COOLER

AIR FOR COOLING

HEATING **D**

CONDENSER

COMPRESSOR

PUMP

HEAT SOURCE

DRAIN

EXP. VALVE

WATER COOLER

HEATING AIR

DRAWN BY HAROLD HERKIMER

INDEX

U

V